What flabbergasted readers are saying about 505 Flabbergasting Facts About Germs!

This book is entertaining ... and scary ... when you realize that the facts are, in fact, the truth!'

I could picture myself in a doctor's office waiting room. I idly pick up [this book] and the next thing you know, I can't put it down and want to schedule another appointment just to finish it!

Wow! I'll never think the same way about germs again! This was more than flabbergasting—it was an eye-opener. Where has all this information been?

Reality TV shows are nothing compared to the real life of germs—our best friends, and our worst enemies. I wonder why the media doesn't cover this stuff more—seems like it's important?

Yuck! Phew! Gross! But real? That's hard to take. Hard-to-believe, but true. I'll never ignore those germy news reports anymore. Sounds like what I don't know could kill me.

I probably never would have wanted to read a book like this until anthrax, flesh-eating disease, SARS, monkeypox, and all sorts of gross germ stuff started appearing in the news. Some of it hit just a little too close to home. Now, I'm fascinated...a germ buff—as long as they stay away from me! A great read!!

Marsh must know more than anyone on the planet about germs! 505 is a fun, funny, but serious and spooky read. Trust no germ is now my motto. I used to make fun of people walking around with their antibiotic sprays and baby wipes—but no more!

No wonder Howard Hughes always washed his hands so much! He wasn't crazy—he was smart. This book is a real eye-opener about the history of germs right up to the present in our own living rooms, bathrooms, baby nursery, backyard, and more. I enjoyed reading it, but now I wonder what's next in germland and is it coming soon to a surface near me?

What people are saying about The Official Guide to Germs...

I love this A-to-Z guide because when I read something in the paper or hear it on the news, I can quickly look it up and at least have some inkling of whether or not I ought to do something about that particular germ. The germ-related info is helpful too. Also, the quickness of use. In the middle of the night, I don't have time to consult a big manual—I just want a clue as to whether I should head on down to the emergency room or not!—Michele Yother, parent of two small children

From my and my business's standpoint I want to learn as much as possible about prevention and avoidance of health-threatening germs. This book provides just that in an easy read, common sense reference guide.—Kathy Gumph, Owner, Allaire Timbers Inn, Breckenridge, Colorado

This guide does a great job of providing simplified information about public health concerns that can be easily understood by the general public. I see the book as a "Cliff Notes" to the public health and hazards surrounding the newest threats and concerns of citizens everywhere. The guide has user-friendly, insightful information ... very useful for any family concerned about the consequences of encountering germs.—Clyde B. Gibbs, Jr., Medical Examiner Specialist, Chapel Hill, North Carolina

It's a germy world and I worry about my new baby, the news I read about E. coli outbreaks, SARS, and other public health threats. This concise, factual, easy-to-read guide helped dispel my fears and help me know when to seek medical advice or further assistance, especially if possibly needed immediately. I have a new baby. What do I actually know about germs? This book is a help.—Michael Longmeyer

505
Flabbergasting
Facts About
GERMS

by Carole Marsh

Copyright © 2003 by Carole Marsh
2nd printing 2004
All rights reserved. No part of this book may be used or
reproduced in any manner whatsoever without written permission
of the Publisher, except for brief excerpts for review.

Library of Congress Cataloging-in-Publication Data has
been applied for.

ISBN: 0-635-01672-9, paperback
Printed and bound in the United States of America

Cover and book design, Victoria DeJoy and Cecil Anderson
Editor, Jenny Corsey

505

Flabbergasting Facts
Coming Soon:

505 Flabbergasting Facts About Sex!

*505 Flabbergasting Facts
About American History!*

505 Flabbergasting Facts About Space!

505 Flabbergasting About Sports!

505 Flabbergasting Facts About Pirates!

505 Flabbergasting Facts About Animals!

Other Germy Books
by Carole Marsh

*Hot Zones!: Viruses, Diseases, and
Epidemics in History+Why to Get Your
Immunizations, Current Biological Events,
and Much More!*
(in an edition for each U.S. state and Canada)

The Official Guide to Germs

*A Kid's Official Guide to Germs:
Our Enemies and Our Friends!*

Table of Contents

"Something funny here."
—Dr. Martin Arrowsmith,
Arrowsmith by Sinclair Lewis

"Wash Your Hands!"

SEVERAL YEARS ago, I wrote a series of books for young readers: *Hot Zones!: Viruses, Diseases, and Epidemics in History+Why to Get Your Immunizations, Current Biological Events, and Much More!* I was inspired by Richard Preston's bestseller, *Hot Zones*, and I thought such a book for kids—with basic background information on germs and such, as well as factual tales from right out their own back door—would be fascinating reading.

What fascinated me was all the information I could not possibly share with kids about germs! Some of it was pretty gross, lots of it was sexual, all of it was intriguing. It was especially enlightening to approach the subject from a state (and Canada) point of view. What I learned is that our media focuses hardly at all (at least at that time) on germs, and when they do, only on the event in their own town or nearby. The (apparently) random meningitis-caused death in a local college student ... the food poisoning outbreak at a local restaurant ... a well-known person who died of AIDS in the nearby big city. Otherwise, in the recent past there has been little or nothing at all in the media to hint that the trail of germs of these seemingly random events are more like a worrisome connected dot-to-dot to death. Only in the last few years has anyone bothered to remind us that germs don't need a passport and your health is only as good as that of the person who sat next to you on your last airplane flight.

Even when some germ news began to break out, who cared? Hantavirus in Arizona? So? Legionnaire's deaths in Philadelphia? Doesn't affect me; I live in Washington state. AIDS? Oh, that's a big city problem. Bubonic plague? You gotta be kidding—that died out a long time ago.

However, through my curiously circuitous chosen method to write about germs, I began to see patterns that made the hairs (or were those germs?) rise up on the nape of my neck. Hmm... The ballet of bacteria ... the virtual virtuosity of viruses ... the pecking order of parasites ... and the fun-loving fungi began to show their true ugly faces to me. Like the elusive Scarlet Pimpernel, he was not here, he was not there, he was not, it appeared, anywhere. No. He—Mr. Germ—was everywhere.

About six feet (taller than I am) of germ books later, the media and the message had caught up with me. Now we all care about germs. But do we care about the right things? As soon as I completed *The Official Guide to Germs*, I was trying to explain what I was trying to do with the book to my artist, a parent of two children.

"You don't want to be an expert," I told her. "No one wants to wade through all these thousand-page-plus books that I have. But you want to be knowledgeable, you want to be aware. You are the one after all who will get up with your feverish husband or kid in the middle of the night and wonder, 'Can't it wait till morning?'" I was on a roll. "Maybe it can, but maybe it can't," I told her. "Most of the time, it is the flu ... but sometimes, seconds count. And when you're at the doctor, you want to be able to ask questions. Sure, he may ask you if you've traveled overseas recently, but maybe he won't, and that might be the key to a quick and effective—even lifesaving diagnosis. And while you're at it," I reminded her fervently, "wash your hands!"

Well, she still pasted up the book for me. The problem was that I only wanted to inform and encourage with that book. So what was I going to do with all the flabbergasting germ outtakes that I didn't dare stuff into that friendly A-to-Z reference?

Yeah, I was the kind of nerdy kid who liked to learn why we have snot, and what makes maggots, and all that greasy, grimy gopher guts and mutilated monkey meat kind of stuff. Only that was gross and funny. Germs are just gross, although any "germologist" will confess that germs have their own kind of beauty. But there's a lot worse than gross. There are smart, determined, mutant, manmade, war weapon germs. As we learned during the 2001 anthrax attacks, they're even coming soon to a mail box (or theater, school, airport, door handle, used hanky, ventilation system, etc.) near you.

Stop, drop and roll ain't gonna get it. Neither is getting under your desk with your hands over your head. Maybe this book should have been subtitled, "How I Learned to Stop Worrying and Love Germs," but I think that title (or close to it) has been used.

Anyway, for better or worse, richer or poorer, in sickness and in health, in Seattle or St. Paul, from Bangor to Biloxi, or from Toronto to Timbuktu, till death do us part, none of us will ever come completely asunder from germs. They were here first; they'll be here last. If humans came from "space stuff" brought here on asteroids, maybe we ourselves are even germs.

So, take a break and sit back and relax and learn about your friends and mine—smallpox, Kuru, Ebola, Yellow Fever, E. coli, the flu, and many more—germs. They're absolutely flabbergasting! (Or is that flabberghasting?!)

Carole Marsh

Carole Marsh
Peachtree City, Georgia

1

Abigail Adams, wife of U.S. President John Adams, spent days scrubbing floors with hot vinegar in an attempt to keep a severe epidemic of typhoid at bay in Braintree, Massachusetts.

2

Abigail Adams once "smoked" any mail received, hoping to kill germs on the letters and packages.

3

Syphilis is on the rise in the United States for the first time in more than a decade. Why?
Gay and bisexual men have let down their guard against sexually transmitted diseases. This follows a previous rate drop each year since 1990. More than 2/3 of the new cases are in men.

4

Savannah, Georgia's first epidemic of Yellow Fever stalked the town in 1820, killing 666 people.

5

Scientists are working to create antibiotics from bat spit, giant leeches, and Komodo dragon saliva.

6

Around 225,000 people die of sepsis
(blood poisoning) each year in the U.S. At least
80% of sepsis patients have something else wrong
with them, are infected, and can die within hours
if not treated. About 1/3 of all sepsis patients die.

7

Scientists have sequenced the complete set of
genes for a microbe that can remove heavy
metals from water. This is called bio-remediation,
the use of plants and microbes to clean up
pollution. The Shewanella bacterium
(found in almost all freshwater sediment)
breathes metal instead of oxygen.

8

In the 1970s, scientists first used
oil-eating bacteria to clean up oil spills.

9

At least 40,000 Americans die annually
as a result of infections gotten from
antibiotic-resistant bacteria.

10

One scientist has figured out how to produce something he calls squalamine, which has antibiotic effects, from ground and cooked shark livers.

11

In October 2002, it was discovered that the West Nile virus could be spread through breast milk.

12

A store called Safer America in Manhattan near Ground Zero sells chemical detection gear, hazardous material suits for the family, gas masks, and other germ biohazard and terrorism-thwarting goods. Check it out at saferamerica.com. Other such stores include Ax-Man Surplus in Minnesota and J&L Self-Defense Products in Berkeley Springs, West Virginia.

13

"Swapping spit" (kissing) boosts the immune system by helping the body beef up its defenses against germs.

14

Most practicing physicians in the U.S. do not know what smallpox looks like because they have never actually seen a case.

15

A "super pox" is a genetically engineered virus that is designed to overcome the existing vaccine against it.

16

Some people mispronounce the common childhood infection impetigo as "infant tigo."

17

Eleanor Roosevelt's mother died of diphtheria.

18

In November 2002, a nationwide class-action lawsuit was filed against Pilgrim's Pride, maker of 27 million pounds of deli meats recalled for possible contamination with listeria, which can cause food poisoning.

19

The two known stocks of smallpox virus are in Atlanta at the Centers for Disease Control and Prevention (CDC) and at the Research Institute for Viral Preparations (also called by the very James Bond-sounding name Vector) in Siberia in Russia. Covert stockpiles of the deadly virus are suspected in Iraq, North Korea, Russia, and France. At last count, Vector had 140 different strains of smallpox. Since 1989 it has been reported that Russia has enough smallpox mounted in intercontinental ballistic missiles and bombs to kill everyone in the world many times over. There have also been doomsday rumors that the Russians have successfully combined smallpox and Ebola into one germ from hell. If true, shall we call it Smallbola or Epox?

20

Unborn babies stock up on disease-preventing immune cells piped from the mother through the umbilical cord.

21

Legionnaires' disease was said to have been found in the hot-water pipes of the Queen's shower at Buckingham Palace.

22

Of the 10 million pounds of antibiotics produced
in the U.S., more than
half goes to livestock.

23

PERVs are Porcine Endogenous RetroViruses.
They can be found in the DNA of pigs.

24

70% of the bacteria that cause infections in
patients in hospitals are resistant to at
least one antibiotic.

25

It takes 17 years to develop
the average antibiotic.

26

VISA infections are vancomycin-
intermediate Staphylococcus aureus
and are extremely drug resistant.

27

A bacterium can develop resistance to an antibiotic in mere minutes.

28

"Be a virus, see the world."
—Gary Larson, cartoonist

29

Researchers today suspect germs
may be the cause of hardening of the arteries,
some diabetes, multiple sclerosis, kidney stones,
and possibly many other diseases or ailments.

30

Dengue fever has also been called "breakbone
fever" and "devil's crunch" to reflect the
excruciating pain the disease produces.

31

Before the development of antibiotics, plague
killed 60-90% of those infected.

32

A 1900 bubonic plague epidemic in San Francisco's Chinatown lasted four years and killed 112 people. The disease had first come ashore via infected stowaways on a ship from Hong Kong. After the 1906 earthquake, the rat population exploded, creating a second epidemic.

33

During a 1793 Yellow Fever epidemic in New York, vigilantes patrolled the streets in an attempt to prevent possibly infected Philadelphia fugitives from slipping into the city.

34

In the summer of 1999, two 11-year-old boys came down with malaria at the Baiting Hollow Boy Scout Camp on Long Island, New York.

35

A 1993 malaria outbreak in Queens was of a deadly strain (Plasmodium falciparum) that kills 3,000 people around the world each day.

36

Because of its many stowaway mosquitoes from West Africa, France is known as the world leader in airport malaria.

37

The term "airport malaria" or "baggage malaria" comes from when infected mosquitoes travel in the cargo holds and wheel wells of jets flying from tropical countries to more temperate ones. The mosquitoes escape and chow down on the first hapless people they encounter!

38

In the 1980s, Ross River virus escaped from Australia to Fiji and Samoa where it started large epidemics of arthritis and fever.

39

Bill Reeves, today an emeritus professor at the University of California-Berkeley, was the first scientist to use dry ice to attract mosquitoes in order to test them for infectious diseases.

40

When the Pan American Health Organization tried to eliminate dengue fever in Central and South America in the 1940s by cleaning up junkyards and tire dumps, the U.S. refused to participate in the program and, therefore, the disease ended up spreading further than ever before.

41

In November 2002, a husband and wife from New Mexico tested positive in New York City for bubonic plague. This deadly disease had not been seen in the city for a century. About half of the 10-20 cases of bubonic plague seen each year in the U.S. come from New Mexico.

42

Bubonic plague is normally transmitted through fleas that feed on infected rodents and cannot be spread person to person. However, in extremely rare cases, bubonic plague can transform into pneumonic plague, a more serious form of the disease that is contagious.

43

In the past 1,500 years, plague
outbreaks have killed about 200 million people.
The most famous epidemic was the Black Death
in Europe which started in 1347 and eventually
killed 25 million people in Europe and 13 million
in the Middle East and China over a five year
period. Some towns lost 90% of their population
to the disease. Bodies lay in homes because there
was no one left to bury them. About 1,000-3,000
people are diagnosed with plague each year. In
the U.S., about one in seven cases is fatal.

44

The West Nile virus in the U.S.
may have originated from an infected farm goose
in Israel, which was probably infected by storks
during their annual migration. Ticks can carry
West Nile virus. In 2003, it was reported that
many more cases of West Nile virus may come
about because of the mild winter season in the
South, which failed to killed mosquitoes.

45

Only about one million doses of
Yellow Fever vaccine are available in the U.S.

46

Biologists believe that the measles virus originally came from distemper, which infects dogs, and rinderpest, a disease of cattle.

47

Biologists suspect that the cause of the common cold, rhinoviruses, originally came from horses.

48

Biologists think the human form of tuberculosis came from a bovine strain of the same disease.

49

Psittacosis is a bacterial disease gotten from parrots.

50

Rocky Mountain spotted fever, a tick-borne infection, has been found in almost every state in the U.S.

51

Bubonic plague can be carried by the fleas of prairie dogs, squirrels, and rodents.

52

In June 2003, more than 50 cases of monkeypox, presumably from infected pet prairie dogs, caused an outbreak in three midwestern states. This was the first instance of monkeypox ever seen outside Africa. First confined to Wisconsin, Illinois and Indiana, the search for infected animals continued to Pennsylvania, Texas, Ohio, and South Carolina after a New Jersey child came down with monkeypox.

Monkeypox is not as serious as smallpox, although it is related to that disease. The smallpox vaccine is effective against monkeypox. Although all the U.S. cases were from animal to human, monkeypox has been known to spread human-to-human in Africa. It was believed that the American outbreak originated from prairie dogs exposed to an infected Gambian giant rat imported from Africa. The animals either came from pet stores or were bought at "swap meets."

Monkeypox Symptoms

- Cough
- Eyes appear cloudy from pus
- Crusty nose
- Swollen lymph nodes in arms or legs
- Bumpy or blister-like rash on the skin
- Fever

Note: Animals afflicted with monkeypox often experience fatigue and/or loss of appetite.

53

In 1998, the Nipah hemorrhagic virus
killed 40% of slaughterhouse workers who
developed encephalitis after being infected by pigs
that had been infected by fruit bats.

54

Leptospirosis is thought to be the most
widespread zoonotic disease in the world. People
who work in rice fields, mines, and sugar
plantations are especially susceptible. At least
100-200 Americans get this disease
each year, usually from fresh water,
wet soil, or vegetation contaminated
by the urine of infected animals.

55

Some scientists believe that the disease
leptospirosis was a factor in Napoleon's defeat in
Poland and Russia during the winter of 1812.

56

More than 16,000 Americans come
down with Lyme disease each year after being
bitten by infected deer ticks.

57

Ehrlichiosis is a disease caused by the same type of deer tick that carries Lyme disease.

58

You can get a Salmonella infection from diseased iguanas.

59

Tularemia, although rare in humans, can be gotten from infected rabbits, cats, and voles.

60

According to the CDC, nearly one in four (at least) Americans are infected each year by something they eat.

61

Alfalfa sprouts can contain Salmonella bacteria.

62

Norwalk virus is the chief cause of foodborne illness in the U.S., with more than 23 million cases per year.

63

Hot dogs can contain the Listeria bacterium.

64

Campylobacter bacteria can be found in turkeys.

65

A 1999 outbreak of Salmonella infections from
unpasteurized orange juice infected
500 and killed one.

66

A 1996 E. coli epidemic stemmed from
unpasteurized apple juice. At least 70 people,
mostly age six and under, were infected. A child
in Colorado died.

67

Food poisoning from some pathogens can kill a
previously healthy person in a week or less.

68

Campylobacter jejuni is the most common
bacteria found in food today.

69

Hemolytic uremic syndrome (HUS) is a sometimes fatal complication of foodborne infection. It causes severe abdominal cramps and bloody diarrhea, which can lead to death, especially in young children and the elderly.

70

Listeria monocytogenes is the most deadly bacteria found in food today.

71

E. coli can infect unpasteurized cider when fallen apples touch deer or cattle waste.

72

In 81% of foodborne illnesses and 64% of deaths, doctors can determine no known pathogen to be the cause.

73

The E. in E. coli stands for German pediatrician Theodore Escherich who first discovered the bacteria in 1885.

74

The letters and numbers in an E. coli
designation, such as E. coli O157:H7 refer to
antigens on the body and flagella (whiplike tails)
of the organism which attack the immune system.

75

The famous Jack in the Box hamburger chain
E. coli epidemic sickened more than 700 people
in 1993 and killed 4 children. The chain's
marketing campaign during this time was
"Monster Burgers—So Good It's Scary."

76

Prior to the Jack in the Box E. coli epidemic, at
least 22 outbreaks of the same form of E. coli had
been documented along with 35 deaths.

77

E. coli 157.H7 contains Shiga toxin, the third most
deadly bacterial toxin after those that cause
tetanus and botulism.

78

In Maine in 1992, a three-year-old boy died of
kidney failure after being infected with E. coli
traced back to organically grown lettuce.

79

Many scientists believe that the now common disinfectant rinses and other methods to kill germs on farms and in food processing may actually help germs become more resistant and develop even more deadly strains.

80

In 1999, 750 teenage girls at a high school drill team camp in Texas became severely ill from E. coli O111:H8, possibly either from chewing on ice or eating at a salad bar.

81

Salmonella infection can be gotten from raw eggs or products containing raw egg. Eggs can be infected after chickens eat feces which can be sucked into the bird's reproductive tract, including the ovaries where the egg is infected even before a shell is formed.

82

Unpasteurized orange juice once caused thousands of Disney World guests to become ill. The juice had been processed at a plant where frogs were infected with Salmonella.

83

A 1998 E. coli outbreak in Indianapolis was traced to the cole slaw in a Kentucky Fried Chicken restaurant. Investigators learned that some of the cabbage used to make the slaw came from farm fields that had been flooded with untreated water where cattle had deposited urine and feces.

84

Microbiologists claim that there is more bacterial contamination on organically grown produce than there is on conventionally grown produce.

85

In Japan in 1986, at least 10,000 people were infected with E. coli from white (daikon) radish sprouts.

86

In 1995, the CDC tracked an international epidemic of Salmonella stanley gotten from eating alfalfa sprouts. The epidemic spanned from Arizona to Michigan to Finland. Possibly 24,000 people were infected.

87

The CDC speculates that most Americans eat Listeria monocytogenes at least once a week, although most don't get sick because some strains may be less virulent than others. However, each year, Listeria sickens more than 2,500 and more than 500 die. Listeria kills 1 in 5 of the people it infects. A 1998 outbreak related to hot dogs killed 15 adults and caused 6 stillbirths or miscarriages.

88

Tests show that up to 18% of food handlers in restaurants have an intestinal infection.

89

The General Accounting Office (GAO) estimates that 85% of foodborne infections come from fruits, vegetables, seafood and cheese.

90

More than 3,000 folks with a sweet tooth became ill with Norwalk virus after an infected baker mixed a batch of buttercream frosting with his bare hands and arms.

91

The FDA inspects less than 1% of food
imported from other countries.

92

As many as 35,000 travelers, in 24
states, the District of Columbia, and four other
countries, became ill with Shigella from tainted
cold meat sandwiches served on airplanes. The
source was a single flight kitchen employee who
had worked with the food while sick
with diarrhea from shigellosis.

93

A federal food safety program, HACCP (Hazard
Analysis Critical Control Point), was developed
from inspections once conducted to protect
astronauts from contaminated food.

94

S.T.O.P is a Chicago-based group promoting Safe
Tables Our Priority. The president is the mother
of a six-year old who died after eating a
hamburger infected with E. coli.

95

Before the invention of antibiotics,
90% of people who got a staph infection of the
bloodstream died. Bacterial pneumonia killed 1/3
of its victims. There was no cure for tuberculosis
or gonorrhea. Rocky Mountain Spotted Fever
killed 20% of those infected. In some parts of the
United States, 10% of women died in childbirth.

96

It is estimated that a single teaspoon of soil may
contain thousands of different species of germs.

97

The Chinese once used spoiled soybean curd to
cure skin infections. Mayan Indians used rotted
roasted green corn to cure skin and
intestinal infections. During the Renaissance,
Europeans let bread mold and mixed it with
water to dress wounds. Today we know that it
was antibacterial toxins, produced by fungi, that
caused these "natural" cures.

98

Antibiotic-resistant germs can survive on
unwashed hands for up to 60 minutes.

99

In 1928, Scottish bacteriologist Alexander Fleming discovered penicillin after leaving a petri dish filled with Staphylococcus aureus sitting around while he was on vacation. When he returned, he found a green mold seemed to have killed off most of the staph germs.

100

The term "hospital germ" was coined in 1949 after the realization that many germs had grown a strong resistance to antibiotics that previously had felled them.

101

The first case of penicillin-resistant pneumococcus was found in 1967 in a healthy three-year-old boy in Papua, New Guinea.

102

In the 1970s, two babies who had attended the same Maryland day care center died from ampicillin-resistant meningitis.

103

A multidrug-resistant strain of tuberculosis killed almost 500 people (mostly AIDS patients) in New York City in the early 1990s.

104

According to the World Health Organization (WHO), 14,000 hospital deaths each year in the U.S. stem from drug resistance.

105

Staphylococcus aureus harmlessly lives on the skin of the nostrils, armpits, or groin of 20-25% of humans. However, this staph germ, from a health care worker or a patient, can get into the body via a urinary or other type of catheter, IV line, or open wound, causing heart valve, blood and bone infections, often leading to death.

106

It is estimated that in some hospitals, at least 70% of staph infections are of a multidrug-resistant strain.

107

During an outbreak of vancomycin-resistant enterococci germs in a Pittsburgh hospital, more than 30 patients in a liver-transplant unit died.

108

Puerperal fever, a type of infection, once killed many women who had just given birth in a hospital. The source of the germ was from the hands of doctors who had performed autopsies in the same clinic and failed to wash their hands before delivering the babies.

109

In 1998, germs beneath the long fingernails of nurses contributed to the deaths of 16 babies in an Oklahoma City hospital.

110

A 1999 Duke University study showed that only 17% of ICU doctors properly wash their hands.

111

Leishmania braziliensis, a microscopic protozoan, eventually rots away the lower half of the face, leaving a person mute and ghoulish-looking.

112

Echinococcus granulosus, a tiny tapeworm, so
entrenches itself in the human brain
and organs that it forms immense cysts
filled with fluid so poisonous that even a small
rupture can cause death.

113

Mummy researchers have found 3,000-year-old
calcified eggs of a parasitic worm, Schistosoma
haematobium, in mummies. These worms so
severely infect their victims' bladders and kidneys
that their urine would have been bright, bloody red.

114

In India and some other countries, a weakened
strain of tuberculosis vaccine is given as a joint
inoculation for both tuberculosis and leprosy. The
connection between the two diseases was made by
an English mummy researcher.

115

Mummy researchers usually pick on the penis for
blood samples of ancient blood because it's more
apt to survive intact. These blood samples allow
scientists to test for blood-borne infections of the
past in order to help better understand how to
deal with them in the present and future.

116

A legend (or is it?) continues to circulate that in the mid-19th century, American mills ran short of rags to make paper. (They had yet to discover the use of wood pulp for this purpose.) As the demand for paper soared, it is said that several East Coast mills imported Egyptian mummies by the shipload and hired local women to strip them of their wrappings. The cloth was then fed into machines which cranked out the type of brown wrapping paper grocers used. According to one paper historian, the Syracuse Standard published an entire issue on sheets of paper made via this process. However, disease from the mummy wrappings was said to have spread in the mills, and the government required all rags to be sterilized to prevent further outbreaks.

117

Megacolon is the name given to a disease instigated by a protozoan known as T. cruzi. Although only as large as two red blood cells, it attacks the nerves of the colon with such a vengeance that the normal contractions of the intestines are stopped. After several months of being unable to defecate, the infected person usually dies of blood poisoning.

118

Infants can get botulinum poisoning from eating honey. (Bees foraging for pollen can pick up spores produced by C. botulinum bacteria that normally live in soil or lake sediments.) Paralysis can be so severe that a baby is unable to open its eyes to cry. The highest incidence of infant botulism in recent years has been in Staten Island, New York. Only a limited amount of antitoxin from humans is produced each year to treat infected infants. It is stored only at the California Department of Health. The antitoxin comes from the blood of people who have worked at the health department and been vaccinated for botulism in the past.

119

AIDS continues to be a major medical problem in many parts of the world. HIV and AIDS are pandemic with more than 34 million people believed to have been infected by the year 2000, resulting in three million deaths annually. In early 2003, the results of the first human trials of a vaccine against AIDS were released. The VAXGen AIDS vaccine proved not to work in general on Caucasians or Hispanics, yet did have a 67% effective rate on African Americans and Asians.

120

AIDS is the leading cause of death of women in South Africa and is claiming increasing numbers of lives each year. At the end of 2002, women made up half of all AIDS cases worldwide.

121

Some doctors are exploring the idea that kids who don't encounter enough flavors of germs in childhood are more apt to develop allergies. Called the "hygiene hypothesis," it is based on the observation that children in developing countries, who constantly fight off infection, rarely develop allergies.

122

Anaerobic bacteria thrive in parts of the body that have very low levels of oxygen, such as the intestine. They also love to cavort in deep or dirty wounds and in decaying tissue. They not only don't need oxygen to live, some forms can't survive at all in the presence of oxygen. While some anaerobic bacteria live harmlessly in our bodies, those that cause infections tend to create abscesses. This includes dental abscesses, jawbone infections, periodontal disease, chronic sinusitis, and middle ear infection, as well as abscesses in the brain, spinal cord, lung, abdomen, liver, uterus, genitals, skin, and blood vessels.

123

Anthrax infections are generally rare in the U.S.; however, in 2001, two ranchhands were infected in Texas and one in North Dakota; all survived.

124

Anthrax occurs naturally in the environment, found in soil. People who work with animal hides are sometimes exposed to anthrax. The country of Afghanistan has soil which contains high amounts of anthrax. Anthrax Island, off the coast of Scotland, got its nickname as the location of explosive testing of anthrax weapons during World War II. After cattle on the mainland caught the disease, the tests were stopped and the island was banned to human visitation for 50 years.

125

It is widely believed that anthrax was the cause of the 5th and 6th plagues in the Book of Exodus in the Bible. As a disease of Roman farm animals, anthrax was described by the poet Virgil. More than a century ago, anthrax was the first disease definitely known to be caused by bacteria. In medieval times anthrax was called the Black Bane for the dead skin lesions that appeared on a person. Anthrax comes from the Greek word for coal, anthrakis.

126

Although offices or buildings contaminated with anthrax can be decontaminated, it may be impossible to completely declare that absolutely no anthrax spores are left. Spores which fall to the ground can no longer cause inhaled anthrax, but could cause the skin form of anthrax.

127

One of the earliest natural antibiotics was Piptoporus betulinus, a fungus of the birch-tree that contains antibacterial oils and was once used to dress wounds. Evidence of this antibiotic was found in a Stone Age mummy.

128

Mary Hunt, a tea lady in the laboratory where penicillin was discovered, brought in a rotting piece of melon that turned out to be the answer to how to grow large quantities of penicillin mold quickly. Following "Moldy Mary's" discovery, the antibiotic went into full production.

129

Bacteria are the dominant forms of life on Earth. There are more germs than any other living organisms—not surprised, are we?

130

How valuable are antibiotics?
During World War I (before antibiotics), one in
five American soldiers died from pneumonia.
During World War II (after antibiotics), less than
1% of soldiers died from this disease. In fact, it
was said that penicillin was actually a factor
in the winning of the war.

131

Babesiosis is an infection of red blood cells
caused by Babesia parasites. This disease is
transmitted via the bite of hard-bodied ticks, the
same deer ticks that cause Lyme disease.
Although common in animals, the disease is rare
in humans. Most cases of this disease are gotten
on offshore islands of New York and
Massachusetts. Most of these cases of babesiosis
are mild. However, for people who have had their
spleen removed, the disease can be fatal.

132

Bio-warfare is not new; it was used in the Middle
Ages (when the heads of people who had died
from disease were catapulted over walls into
cities to infect citizens) and in early America (via
"gifts" of smallpox-infected blankets) to wipe out
entire tribes of Native Americans.

133

Ascariasis is an infection caused by an intestinal roundworm. This disease is most common in warm areas with poor sanitation, but does occur worldwide. The larvae travel from the intestine through the bloodstream to the lungs, where they climb up the respiratory tract and are swallowed, return to the intestines, and start the cycle over again. The larvae mature in the small intestine, where as adult worms they live and grow to be 6-20 inches (15.24-50.8 cm) long. Symptoms can include fever, coughing, wheezing, and abdominal cramps. A heavy concentration of worms can cause poor absorption of nutrients by the body, and can obstruct the appendix. Good sanitation and the avoidance of contaminated vegetables are the best methods of prevention.

134

In the 1950s and 1960s, during the Cold War, volunteers in Operation Whitecoat were intentionally infected with potential bio-warfare diseases; all were monitored, treated, and recovered.

135

In 1977, jalapeno peppers were the cause of botulism illness in 27 customers of a Michigan restaurant.

136

Bacteriophages, also just called phages, may offer medical hope and help that rival that of antibiotics when they were first discovered. Bacteria can be killed by viruses. These viruses (phages) may serve as a living, natural antibiotic against disease. There are many different types of phages; each is designed to kill only certain kinds of bacteria. Phages actually infect bacteria. Because phages multiply incredibly rapidly, they offer bacteria little time and few ways to defend themselves. Phages to be used in disease treatment were known but ignored when antibiotics first came along. But as scientists stew and worry over the increase of germ resistance to many (and sometimes all) forms of antibiotics, they are re-exploring phages and their future usefulness as germ fighters. Phages have been successfully used to fight numerous diseases. Experiments have injected phages directly into the parts of the body under siege by disease. In some tests, phages were more effective than many antibiotics. Whether or not phages turn new pages in the common treatment of disease remains to be seen. However, phages are everywhere you find bacteria (although they are only 1/40 the size of bacteria.) One milliliter of seawater can contain one million phages.

137

In 1863, a Confederate surgeon was
arrested and charged with trying to import
Yellow Fever-infected clothing across Union lines
during the American Civil War.

138

World War I veterans who suffered from the
effects of mustard gas settled in
Twentynine Palms, California, in hopes of
benefiting from the dry desert air.

139

A NATO handbook lists 31 infectious agents as
potential biological weapons; it is believed the top
four are anthrax, smallpox, plague and botulism.

140

After eating patty melts at a restaurant in
Peoria, Illinois in 1983, 28 people fell ill from
botulinum on onions cooked and held
at room temperature, which were not
reheated before being served. While
no one died, almost half of the victims
required breathing assistance at a hospital.

141

There were 74 outbreaks of botulism
in the U.S. between 1983 and 1987, infecting
140 people and killing 10.

142

Blast it! Who wants blastomycosis? This infection
is caused by a fungus and produces a disease also
known as Gilchrist's disease. While usually a lung
infection, the disease can spread through the
bloodstream to other parts of the body. Most
blastomycosis infections occur in the U.S.,
especially in the Southeast and Mississippi River
Valley. Although it is not known where the fungal
spores originate in the environment, beaver dams
were associated with one outbreak of the
disease. Most affected are men between the ages
of 20 and 40. Symptoms include fever, chills,
excessive sweating, cough, chest pain, and
breathing difficulty. Prompt treatment can avoid
more serious health problems (infection of the
prostate, for example), and even death.

143

Botulinum is 6 million times more toxic than
rattlesnake venom. A lethal dose is a mere
1/10,000th of a milligram.

144

Each year about 100 infants develop botulinum poisoning from spores in dust or in the air. The spores germinate in the baby's gut, churning out toxin. A limited amount of human antidote (called BIG for botulism immune globulin) is stockpiled by the California Department of Health to treat such infants each year. Infant botulism is the most common form of botulinum poisoning in the U.S. today. (Adult botulism only occurs in around 10 people each year, gotten from contaminated food, dirty needles, or infected wounds. Adults are usually treated with an antitoxin made from horse plasma.)

145

Cervical cancer is much more likely to strike a woman who is already infected with human papilloma virus (HPV)—the one you get your "Pap" smear to detect—so be sure to do that annually. Scientists also think that this virus is connected to some types of mouth cancer.

146

Don't tell Starbucks: Researchers have proven that drinking tea boosts the immune system, which then responds five times faster to germs, than after drinking coffee.

147

Bovine Spongiform Encephalopathy (BSE) may be caused by eating infected beef. This is the infamous "mad cow" disease that ravaged cattle in England in the last few years, necessitating the killing of entire herds. A number of people contracted the disease and died. There is no cure. The disease destroys brain cells, leaving the tissue with a spongelike (or Swiss cheese hole) pattern. This destruction causes the breakdown of all physical and mental functions. For a while there was some fear that the disease would spread to other countries, even across the Atlantic Ocean to the United States. Farmers took great precautions to avoid such a spread of the disease. BSE may be related to a similar disease called New Variant Creutzfeldt-Jakob Disease (CJD).

148

Botox, the popular "cure" for facial wrinkles, is made from a form of botulinum.

149

Cat-scratch disease is an infection at the place where a cat, that has the bacterium Bartonella henselaw, has scratched you. Although the cat may show no signs of illness, the bacteria can infect the walls of your blood vessels.

150

In April 2002, scientists reported preliminary success with using a virus called ONYX-015 to kill cancer cells (while not killing healthy cells), offering hope for an additional tool against cancer. By 2003, scientists had successfully tested a vaccine against cervical cancer. If additional tests are successful, doctors hope that this second most deadly cancer for women (breast cancer is first) might be ended worldwide in the future. It is possible that young men, who can spread the HPV virus, might also be vaccinated. It is also believed that the vaccine might offer protection against some types of cancers of the genitals, anus, head, and neck, as well as a rare form of HPV infection which affects children born to mothers with genital warts.

151

More than 50% of untreated people with severe cholera die; fewer than 1% die who have prompt treatment to replace body fluids.

152

In 2003, it was determined that a simple water filter made from cloth (such as an old sari) could reduce cholera cases by half.

153

Should "germs" ever be listed at the "cause of death" on a death certificate? You bet! Germs are often the unofficial—but real—cause of death. Deaths, especially as a result of infections gotten during a hospital stay, are often hidden in other categories. Yes, perhaps a person did actually die of heart failure . . . but it may have been brought on by germs. According to the CDC, in 1999 of 10 million patients entering U.S. hospitals, 2 million caught bacterial and viral infections from which 90,000 died. This makes hospital infections the number five killer in the U.S. Nearly half of these deaths result from bacteria, usually S. aureus.

CAUSES OF DEATH (U.S. 1999)

1.	Heart Disease	724,621
2.	Cancer	549,761
3.	Stroke	167,261
4.	Chronic lower-respiratory disease	124,141
5.	HOSPITAL INFECTIONS	90,000
6.	Unintentional injury	86,909
7.	Diabetes	68,394
8.	Influenza and pneumonia	63,408
9.	Alzheimer's Disease	44,536
10.	Nephritis	35,359
11.	Septicemia	30,397

154

In April 2002, researchers reported that genetically tweaking bacteria normally present in the mouth could create a lifelong immunity against cavities. Human trials using the modified streptococcus mutans bacteria may begin in 2003. If the scientists pull this one off, everyone will be smiling ... except perhaps dentists!

155

An extreme form of the bacterial infection, cellulitis, is necrotizing fasciitis, also known as the "flesh-eating" disease. Caused by a nasty strain of the strep germ, it can destroy tissue rapidly. The affected skin may turn violet and large fluid-filled blisters, and even gangrene, can develop. Severe cases require amputation, and the death rate is about 30%.

156

Non-cholera vibrio infections are reported each year, around half from Gulf Coast states, and the remainder from numerous other states. Most of these infections occur during summer months and are related to the consumption of contaminated seafood.

157

The disease Chagas is closely associated with poverty. The blood-sucking triatomine bug which transmits the parasite often lives in the walls of urban slums or rural homes. It can also be transmitted via infected blood. Following infection, the disease may be dormant for several years, but irreversibly affect internal organs, especially the heart, esophagus, colon and nervous system. After years of no symptoms, an infected person may develop nerve problems, digestive troubles, or cardiac symptoms which can lead to sudden death. Chagas disease is more widespread in Latin America than HIV. It infects 16-18 million people every year, killing more than 40,000. Scientists report that hundreds of thousands of Latin American immigrants in America are infected with Chagas. In the American South, assassin bugs can spread Chagas. A 1970 test showed that one in every 40 Texans tested positive. A 17-year-old boy died from heart inflammation after receiving blood tainted with Chagas.

158

The term "vaccination" was named after vacca, the Latin word for cow.

159

The varicella zoster virus that causes chickenpox stays in the body for life, once you have been infected. While you can't get chickenpox again, this same virus also causes "shingles." An adult with herpes zoster can infect a person who has never had chickenpox with that disease. In 2003, the chickenpox vaccine showed that it could help prevent or allay an outbreak of shingles in adults, which can be very painful. Also in 2003, late-night talk show host David Letterman missed several shows due to a bad bout of shingles.

160

The first major cholera epidemic occurred in India in the early 1800s. Cholera reached the United States in 1832 via sailors and immigrants. The famous 1854 outbreak in London was stopped when traced to a contaminated public water drinking pump. Major cholera outbreaks have occurred in the U.S. and Canada. A 1991 epidemic in Peru spread throughout Latin America, killing 11,000.

161

When you sneeze—germs can shoot out at a race car pace of 100 miles
(161 km) per hour!

162

In early 2003, a 58-year-old Summerville, South Carolina man was diagnosed with Creutzfeldt-Jakob disease. While doctors suspected that he had been stricken with the classic form of this rare disease, it was also possible that he had the new variant. In the 1980s, the man had traveled in Scotland, known as a source for the new variant form of the disease. After suffering what one doctor described as "Alzheimer's on fast forward," the man died.

163

Clostridia produces toxins that damage tissue or the nervous system. Most such infections are the result of mild food poisoning. However, one form of clostridia, necrotizing enteritis, can cause inflammation that can destroy the walls of the intestines. Clostridia can also come from eating contaminated meat. Botulism can come about from eating food contaminated with a toxin produced by some clostridia. Clostridia can also cause gangrene and tetanus.

164

Many flu epidemics start in Southeast Asia, from human contact with infected pigs, ducks, or chickens.

165

Cytomegalovirus (CMV) is a common virus that eventually infects us all, without any symptoms. A pregnant women can pass CMV to her fetus, where it can cause hearing loss, mental retardation, and other birth defects.

166

New Variant Creutzfeldt-Jakob Disease (CJD) is a disease which destroys brain cells. It may be related to BSE (Bovine Spongiform Encephalopathy), or "mad cow" disease.
The disease is fatal. Although it occurs throughout the world, little is known about how it spreads. The disease primarily affects adults, especially in their late 50s. A similar disease (called scrapie) affects sheep and cattle (mad cow disease). It is believed humans can get the disease from eating infected meat. Once infected, symptoms can occur months or even years later. Symptoms include brain damage and dementia (the loss of intellectual ability). Once the symptoms begin, the disease progresses rapidly. There is no cure. The disease once could not be confirmed except by an autopsy of the brain. However, a test for the disease is now possible.
Note: Jakob is pronounced YAH-cob.

167

Only one person in a million worldwide gets the
original form of Cruetzfeldt-Jakob disease. There
have been at least 100 victims of the new variant
form related to mad cow disease. To date,
no case has ever been confirmed that
originated in the United States.

168

To help reduce chances that an infectious form of
Cruetzfeldt-Jakob disease could become
established in America, the government is
considering a ban on "high risk" animal parts used
in some products. This includes the brains, spinal
cords, and some other animal parts leftover
from the slaughtering process which are
then put into feed for other animals.

169

In April 1993, in Milwaukee, Wisconsin, untreated
water from a spring contaminated the local
drinking water with Cryptosporidiosis, a water-
borne protozoa. Of 800,000 users of this water
supply, almost half became ill; 4,400 were
hospitalized, and 40 died.

170

Coccidioidomycosis, also known as "San Joaquin fever" or "valley fever," is an infection, usually of the lungs, caused by a fungus. Symptoms may be mild, occurring 1-3 weeks after infection and require no treatment, or can be more severe (perhaps not showing up until after you have left the area where you became infected, travelers!), with chills, fever, and coughing up blood. Some people develop desert rheumatism, which includes eye inflammation, arthritis, and the formation of nodules on the skin. A more progressive form is unusual but can cause severe health problems in the lungs and other parts of the body over time. No treatment may be required for mild cases, while more serious cases require antibiotics and even long-term treatment of injections into the spinal fluid.

171

Cowpox is a disease that affects cattle. It is similar to smallpox, only milder. In 1796, Edward Jenner, an English doctor, gave the first successful vaccination, against smallpox, to an eight-year-old boy. Jenner had noticed that the farmers and milkmaids who caught cowpox seemed to be immune to smallpox. He used fluid from a cowpox blister to develop a mild form of smallpox to create his successful "vaccination."

172

Vacation From Hell?: The CDC claims that despite the widespread media coverage of cruise ship sickness outbreaks, the likelihood of catching a germ at sea is actually going down, not up—that there are just more ships and passengers and cruises out there than ever before. That's probably little comfort to cruisers who find themselves suffering from nausea, vomiting, and diarrhea for all or part of their vacation ... or to cruise ship operators being slapped with lawsuits.

173

Even in the midst of the anthrax attacks of late 2001 (when 5 people died and 18 were infected), the CDC was also busy coping with an epidemic of dengue fever in East Maui, Hawaii. This disease is endemic in at least 100 countries. There are 50 million cases of dengue infection each year. Dengue has also been called "breakbone fever" and "devil's crunch" in reflection of the excruciating pain the disease produces. In 1999 and 2000, more than 200 Americans were suspected of having dengue fever. Those actually diagnosed with the disease had traveled to tropical locales such as Asia, the Caribbean, Central and South America, and Africa, where this mosquito-borne disease is common.

174

In the fall of 2001, 16 children contracted E. coli from cattle at a Pennsylvania petting zoo. In January 2002, one child, a four-year-old who suffered severe kidney damage and had most of her colon removed, received a kidney transplant from her father.

175

What's the difference between diarrhea and dysentery? Dysentery is a serious form of diarrhea. Amoebic dysentery is common in some parts of the world, where it kills millions of children every year. During World War II, soldiers suffered and even died from dysentery. In fact, during some wars, diseases killed more soldiers than combat. Today, treatment can include rehydration via a powder of sodium, potassium salts, and sugar. Kits of such medicine have saved many lives around the world.

176

It is possible that George Washington died of diphtheria.

177

Orchitis is a bacterial infection of the testes.

178

The bacteria that cause diphtheria never actually enter the body. Diphtheria is inhaled and makes itself at home in the throat where it releases a toxin that causes a grayish film on the throat. The toxin can be released into the bloodstream. In severe infections, diphtheria can create an irregular heartbeat, difficulty swallowing, and coma. The disease is usually spread through contact with saliva or nasal secretions of an infected person.

179

Bio-Fashion! What's the difference between the clunky "blue suit" and the clunky "orange suit" you see biohazard workers wear? The "Blue Suit" (yes, they do call it the "blue suit") is really a Chemturion pressurized, heavy-duty biological space suit. It is worn in Biosafety Level 4 laboratories to conduct work on deadly viruses. The "Orange Suit" is a portable, pressurized space suit that has a battery-powered air supply. It is used in the field when working where biohazards may be airborne.

180

A human bowel movement contains 100 million million million bacteria.

181

Erythema infectiosum, nicknamed "slapped cheek disease" and caused by parovirus B19, mostly affects children. It is also commonly called Fifth Disease. Yes, there is a Sixth Disease—Roseola, a common disease of children!

182

Earaches aren't child's play; check it out! Herpes zoster of the ear, also known as Ramsay Hunt's syndrome, is an infection of the auditory nerve. The Herpes zoster virus infects this nerve, causing severe ear pain, hearing loss, and vertigo. Facial nerves can also be infected, causing temporary or permanent paralysis. Hearing loss is possible. Vertigo can last several weeks. The infection is usually treated with the antiviral drug acyclovir. Surgery may be required. Other germy ear ailments include infection of the ear canal, infection of the cartilage of the outer ear, inflammation of the eardrum, infection of the middle ear, infection of the bone behind the ear, and others.

183

Polio is caused by a virus that can only live in humans.

184

Is There an Echo 13 in Here? Echovirus type 13, once considered rare in the U.S., showed up in greater numbers in 2001, causing 76 cases of meningitis in 13 states, possibly triggering outbreaks. States with the largest numbers of reported Echovirus 13 cases included Louisiana, Mississippi, Montana, and Tennessee. In 2000, Echovirus 13 was also first reported with outbreaks in England, Wales, and Germany. This virus tends to infect children, usually in summer and fall. The only recommendation for avoidance of this disease is using good handwashing measures, especially after changing diapers, and other general hygiene practices.

185

As many as one million Americans have HIV infections, with about half undiagnosed or untreated.

186

A disease called "farmer's lung" comes from inhaling the spores of moldy hay. Aflatoxins come from fungi that grows on peanuts. The fungus Stachybotrys chartarum lives in damp wallpaper and wet building materials that contain cellulose.

187

Encephalitis is a serious viral disease, usually gotten from the bite of a mosquito that has picked up the virus from infected birds or pigs. Symptoms include fever, drowsiness, and can lead to brain damage. There are often major epidemics of encephalitis in parts of Asia where there are many water-sopped rice fields. A worldwide epidemic of encephalitis in 1920 killed many people and sent others into comas. The drug levodopa brought some comatose patients "back to life" but only for a short time. The most common forms of viral encephalitis in the U.S., transmitted by insect bites, are:

- Western equine: occurs throughout the U.S.; mostly affects children under one year of age

- Eastern equine: more likely to be fatal; affects young children and adults over age 55

- St. Louis: outbreaks have occurred throughout the U.S., especially in Texas and the Midwest

- California: affects mainly children; includes the viruses California (western U.S.), La Crosse (Midwest), and Jamestown Canyon (New York westward)

188

In January 2002, the American Academy of Pediatrics and the CDC recommended that all newborns be vaccinated against Hepatitis B before leaving the hospital.

189

The flu pandemic of 1918-1920 was one of the worst epidemics to ever occur in America. It began in Kansas, then swept around the world, killing 20 million, compared to World War I's 15 million combat deaths.

190

In early 2003, studies reported that inspections of chicken sold in supermarkets showed that half of the chickens tested were contaminated with bacteria dangerous to human health. Tests also showed that up to 90% of the bacteria were resistant to common antibiotics.

191

During the time of the Black Death, Pope Gregory VII began the habit of saying, "God bless you" to those who suffered from excessive sneezing as a result of the plague.

192

Glanders is an infectious disease
caused by the bacterium Burkholderia mallei. It
primarily affects horses, donkeys, mules, goats, and
sometimes, dogs and cats. While there has been
no report of a case of glanders in humans since
1945, it does sometimes occur among lab
workers and people such as veterinarians who
are in long-term direct contact with infected,
domestic animals. Human-to-human transmission
has also been reported. The disease has been
considered as a possible bio-warfare agent. There
is no vaccine for glanders, which is common in
animals in Africa, Asia, the Middle East, and
Central and South America.

193

Hepatitis C is usually transmitted via infected
blood; there is no vaccine. Symptoms include
fever, jaundice, and a general feeling of sickness.
Treatment with drugs is available, but most
people will carry the virus for the rest of their
lives. This disease can incubate for decades.
Hepatitis C has greatly increased in
recent years. Some scientists estimate
that by 2010, it may strike down more
Americans each year than AIDS.

194

Hantavirus is a disease that has only appeared in recent years. Between 1951 and 1954, more than 3,000 United Nations forces in Korea were infected with this "hemorrhagic fever with renal syndrome," as it is known. Hantaviruses live in mice; people can be infected via infected mites or inhale infected dried urine from the mice. In 1993, a version of the hantavirus that produces rapid respiratory failure showed up in the Four Corners region of New Mexico, Arizona, Colorado, and Utah. Within a few days, a young, healthy Navajo woman died of the disease, as did her boyfriend a few days later. In 1998, 30 people got hantavirus pulmonary syndrome (HPS) in 12 states; nine died. HPS is not contagious from person to person in the United States.

195

One out of every two people in low-income countries dies at an early age from infectious diseases. Most of these deaths are preventable. Such diseases kill at least 17 million people each year; one-third of these diseases occur in developed nations.

196

Coronary heart disease is the leading killer in the U.S. While it is generally related to smoking, lack of exercise, and a fatty diet, some studies indicate an association between infected gums and heart disease. In tests of plaque removed from clogged arteries, 72% contained oral bacteria. Such bacteria in the mouth can easily enter the bloodstream (and therefore the heart) via diseased gums. Therefore, doctors suspect that treating gum diseases such as gingivitis (half of all Americans over age 30 have this) and periodontitis (1/3 of adults over age 30 have this), may help prevent heart attacks. Links between periodontal infection and other illnesses such as diabetes, chronic respiratory disease, stroke, and low birth weight have also been noted.

197

Six diseases account for 90% of the deaths from infectious disease worldwide: influenza; HIV/AIDS; diarrheal diseases; tuberculosis; malaria; measles.

198

In 2003, scientists declared that secondhand smoke leads to cavities in children.

199

One source of infection is medical
devices left in the body. Instead of adhering to
cells as they usually do, germs may attach
themselves to catheters, artificial valves, or other
medical devices and begin to form colonies, which
can spread and cause disease.

200

Even in modern times, isolated groups
such as the Inuit people of Alaska have suffered
from common diseases (such as measles),which
can kill off large numbers of people unless
immunity is established through a thorough
immunization program.

201

Invasive candidiasis is a fungal infection
that occurs when Candida enter the blood.
This yeast fungus causes an infection of the
bloodstream which spreads throughout the
body. Candidemia, one form of this disease, is the
4th most common bloodstream infection among
hospitalized patients in the United States.
Those especially at risk are low-birth-weight
babies, surgical patients, and those whose
immune systems are deficient.

202

Leishmaniasis is a skin disease that is common in people living in refugee camps. In late spring 2002, a serious outbreak was reported in war-torn Afghanistan. The disease can leave victims severely scarred and stigmatized for life, especially girls and women in certain countries, who find themselves so ostracized that they can never even marry.

203

Kuru is a disease that exists at this time only among the Fore people of Papua, New Guinea. From about 1950 to 1970, more than 3,000 of them died from a "laughing disease," as they called it. The Fore are cannibals. They honored their dead by eating them, especially the brains, considered a delicacy. The women of the tribe served as chefs for these meals, explaining why eight times more women died from this disease than men. Scientists learned that the cause of the disease was a prion, a single protein that can cause infection. Because prions are a natural part of the body, the immune system does not recognize them as dangerous. Therefore, they can lurk in the body, only to cause a sudden deadly disease in the future.

204

According to a news report, the Legionnaires' bacteria was found in the hot water pipes of the Queen's shower at Buckingham Palace.

205

Legionnaire's Disease got its unusual name because of its association with The American Legion military veterans' organization. In 1976, they met in Philadelphia. Many members staying in the same hotel came down sick with a flulike disease that turned into pneumonia. Several people died. The bacteria that cause this disease thrive in water of a certain temperature, but must also be misted in order to get into human lungs. Therefore, cooling towers, whirlpool spas, and such offer the best opportunities for potential danger.

206

Most people carry leprosy antibodies because we have been exposed to the bacterium at sometime during our lives.

207

About 5,000 people in the U.S. have leprosy; most are immigrants from developing countries who live in California, Hawaii, and Texas.

208

Leprosy (also called Hansen's disease)
is an example of an "ancient" disease that in
actuality is still with us, even if rare and remote.
Leprosy had its horrible heyday in the Middle
Ages and is believed to be one of the oldest of
human diseases. It is caused by Mycobacterium
leprae and most likely spreads through infected
nasal mucus. Leprosy is only infectious during its
early stages; fortunately, it is not particularly
contagious, nor is it hereditary as was once
believed. Although leprosy is rare in Western
nations, there are still about six million infected
people around the world today. Once cast
out and isolated from society, patients with
such advanced cases that they can no longer
live independently are cared for in modern
leprosy hospitals in countries where the
condition is more common.

209

A 1981 outbreak of listeriosis in
Canada was tied to coleslaw prepared with
contaminated cabbage.

210

In 2000, Lyme disease reached a record high, with
17, 730 cases reported, up 8% from 1999.

211

Body lice are not the same as head lice;
they are found on the body,
and in clothing and bedding.
Crab lice infect the pubic area.

212

Leptospirosis is a group of infections caused by
the Leptospira bacteria. These infections include
Weil's syndrome, infectious jaundice, and canicola
fever. This disease occurs mostly in animals, both
wild and domestic. People get the disease
through exposure to infected animals or their
urine. Farmers, sewer workers, and those
who work in slaughterhouses primarily get the
disease. However, about 40-100 cases of other
people infected with the disease occur each year
in the U.S., usually in late summer and early fall.
They have most likely gotten the bacteria by
swimming in contaminated water.

213

In 1998, BilMar Foods (parent company,
Sara Lee), a processing plant in Zeeland, Michigan,
voluntarily recalled millions of packages of hot
dogs and lunch meats after a listeriosis outbreak
was linked to its products. More than 100 people
were infected; 21 died.

214

Listeriosis is a type of food poisoning.
It comes from food contaminated with the listeria
bacteria. This bacteria is found in livestock and
soil; it can also be found in unpasteurized dairy
products, on fresh vegetables, in shrimp, coleslaw,
and paté. Unlike most food-borne pathogens,
listeria can multiply in the refrigerator. While
most germs are prevented from passing through
the placenta and attacking a fetus, listeria is able
to do so. Listeria sickens about 2,500 people and
kills about 500 each year in the U.S.

215

In September 2002, at least 13 people died of
Listeria infections in Pennsylvania, New York, New
Jersey, and Connecticut. Although
contaminated food was the likely cause, no
specific source was ever determined.

216

In late 2002, more than 27 million pounds of
turkey and chicken sandwich meats became the
largest ever recall of food to be tested
for listeria contamination.

217

Lyme disease was named after a 1975 outbreak in Old Lyme, Connecticut, where a number of children were infected. It is believed that the disease has been around for a much longer time, but has just gone undiagnosed. Lyme disease has now occurred in 47 states.

218

The Machupo virus is so rare that it has only ever been seen in one epidemic site, in Bolivia. Its source is in mouse urine. An infected person finds himself in and out of consciousness and so sensitive to touch that he cannot stand a sheet over him. Blood leaks into the eyes turning them bright red, and small bits of fluid leak out of microscopic holes all over the person's skin. The death rate from this rare deadly disease is 50%.

219

When my father was stationed with the U.S. Army in the Pacific during World War II, he contracted malaria. Many soldiers did. Malaria is caused by a single-cell organism contracted through the bite of an infected mosquito. Today, malaria is often brought to new destinations via water in the wheel wells of airplanes, for example.

220

At least 500 million cases of malaria are recorded worldwide each year, with more than a million deaths. Quinine, from the bark of chinchona trees, has long been used to treat malaria. Because quinine has such a bitter taste, it is said that the British created the "gin and tonic" to help the medicine go down a little better. Today, drugs such as chloroquine are more likely to be used to combat the disease. Some tourists headed for a malaria-prone country take such drugs up to a couple of weeks ahead of time in an attempt to ward off infection. However, resistance to some malaria drugs has become so bad in some nations that it has unfortunately virtually done away with the cheapest form of medicinal protection. Multi-drug resistance has been noted in some places. To beat the bugs at their own game, some places have tried using insecticides on mosquito netting which has cut the infections of babies by half. While this seems like a good thing, it also prevents children from developing an immunity, leaving them susceptible to more severe illness as adolescents.

221

In rural Africa, babies may be bitten by so many mosquitoes each night that they suffer severe anemia.

222

Malaria comes from a term for bad (mal)
air and was long associated with diseases that
seemed to come from smelly summer swamps,
such as coastal Carolina rice plantations during
the colonial era of American history.

223

In 2000, Bill Gates, CEO of Microsoft,
and his wife Melinda, donated $40 million
toward efforts to reduce the global incidences
of malaria by 50% by the year 2010.

224

If you are traveling to a part of the world
where malaria is endemic (Asia and Africa),
you should start preventive treatments at least
a couple of weeks before your trip.

225

Some researchers fear that global warming
will increase the incidence of malaria.

226

In February 2002, a New Jersey woman
died of a form of meningitis which strikes around
3,000 Americans each year, killing 300.

227

Scientists have hopes that artificial genetically-created germs can help mankind. One such new germ life form in the works is based on a parasitic bacterium called Mycoplasma genitalium. This simple organism, which has only 515 genes, lives in the human reproductive tract where it sometimes causes inflammatory infections. Scientists believe that they can genetically alter the germ, substituting artificial genes for real ones. Then, "sparked" by electricity, they think they will end up with a new single-cell life form that may be used to make clean hydrogen fuel, or, perhaps, break down greenhouse gas or other useful tasks that help mankind.

228

Marburg is a viral disease. It first appeared in 1967 in laboratory workers in Marburg, Germany and Yugoslavia. The technicians had been working with cells from African grassland monkeys. Little is known about Marburg. Its symptoms are similar to Ebola—high fever, rashes, and bleeding internally and externally. It is contagious, through blood, breathing, semen, and urine. An antiviral drug may help if treatment is begun early, but the death rate is still high from this disease.

229

In 1492, Columbus and his crew brought smallpox and measles to non-immune native peoples in the Americas. Measles became one of the common so-called "childhood diseases." Today, measles is rare in developed countries due to widespread immunization programs.

230

One of the most common diseases affecting the brain is meningitis. It is caused by a virus or bacterium. Meningitis is named for the meninges, the three membranes that protect the brain and spinal cord. When infected, the meninges can swell and press painfully against the skull. Untreated meningitis can lead to brain damage or even death. The disease is fatal in 10-13% of cases; 10% of those who survive the disease have severe impairments such as mental retardation, hearing loss, or loss of limbs. Those at highest risk are babies under the age of one year, the elderly, and students ages 14-22.

231

During a 1974 meningitis outbreak in Brazil, 3 million people were vaccinated in 5 days to put an end to the epidemic before it could spread across South America and beyond.

232

By late 2002, 14 states had laws requiring proof of vaccination against bacterial meningitis for students entering colleges. These states include Maryland, Virginia, Pennsylvania, Connecticut, Florida, California, Delaware, New Jersey, Arkansas, Texas, Illinois, Michigan, Indiana, and South Carolina. This followed a 1999 recommendation by the CDC after the deaths of several college students in Maryland. First-year college students living in dorms have as much as a sixfold increase in risk of getting the disease as others their age.

233

Infectious mononucleosis (also referred to as "mono") is caused by the Epstein-Barr virus (EBV). EBV is believed to be present in saliva. When infected, young children may show no symptoms. Older children and adults may have fever, fatigue, enlarged neck lymph nodes, and inflamed throat and tonsils.

234

Before the age of five, around 50% of all American children have had an Epstein-Barr virus infection.

235

In 1985, the tiger mosquito came to the U.S. in water-logged Asian tires. Within two years, these carriers of yellow and dengue fevers, as well as other diseases, had spread to 17 states.

236

Under optimum conditions, some germs can multiply every 20 minutes. As your most math-savvy child can tell you, billions of bacteria can be created in just a matter of hours.

237

Papaya, also known as the Paw Paw plant, is used by some people to expel intestinal worms, stop dysentery, and to treat some bacterial infections.

238

Mumps is caused by the mumps virus. One of the more painful, older childhood diseases, mumps is now preventable with a vaccine. If you were born before 1957, you are considered to be immune. The MMR (measles, mumps, rubella) vaccine is recommended for children, and adults who are in college, work in a hospital, travel internationally or on a cruise ship, or are women of child-bearing age.

239

Mycobacteria come in many forms.
One type can cause tuberculosis. Other types
cause much less serious infections. Although
mycobacteria are common (they can grow in
swimming pools and home aquariums, for
instance), they usually only affect people with an
impaired immune system. Mycobacteria can
infect the lungs, lymph nodes, bones, skin, and
other tissues. Some infections clear up without
any treatment. Other infections require months
of antibiotics. One type of this germ can infect
artificial body parts such as breast implants or
mechanical heart valves.

240

"Hannibal the Cannibal" at least knew what to
eat that was good for him. Fava beans are known
to offer antimalarial benefits.

241

In 1995, the nerve gas sarin was
intentionally sprayed in subway trains in Tokyo,
Japan. Sarin is colorless and odorless.
Passengers fell to the ground as their nerves
failed and they could not breathe; some lost
control of their bowels and bladders as
nerve control was lost; a few people died.

242

In 27 weeks, from September 1918 to
March 1919, more than 6,000 people died of
Spanish Influenza in Boston; 14,014 in Chicago; 15,
785 in Philadelphia; and in New Orleans, more
than 33,000 died of the disease, which went on to
spread around the world, especially via infected
military personnel heading off to World War I.

243

Castanospermine, a poison (also known as the
Moreton Bay Chestnut), can be used to fight HIV.

244

In AD 542, Justinian's plague in Egypt, became the
first known pandemic. As many as 100 million
people died (possibly 10,000 per day
at the peak of the pandemic), contributing to the
fall of the Roman Empire.

245

In 2002, doctors reported piercing problems for 1
in 5 students in one New York college. Bacterial
infection was the most common problem,
followed by bleeding, and injury or tearing at the
site of the piercing.

246

In 2002, it was reported that more than 100 California women developed nasty boils and skin ulcers after using a whirlpool footbath that was a breeding ground for the mycobacterium fortuitum microbe. Many cases turned out to be resistant to antibiotics. Other outbreaks may be occurring around the country, but many go unrecognized because they take so long to appear.

247

Pinworms are tiny, parasitic worms that live in the large intestine. The female worms lay their eggs around the anus at night. If you have a case of pinworms in the family, it can quickly spread, and you'd better stock up on detergent because a whole lotta washing will be going on!

248

In November 2002, a husband and wife from New Mexico tested positive in New York City for bubonic plague, the first cases seen there in a century. Each year, there are 10-20 cases in the U.S., half from New Mexico; one in seven die. Worldwide, 1,000-3,000 cases occur each year.

249

Of piercing concern are recent reports by doctors at Yale University of the source of a brain abscess in a young woman following an infection of her pierced tongue. About a month after having her tongue pierced, it became swollen with infection. She also experienced a foul-tasting discharge. The woman removed the jewelry and the infection cleared up. However, in another month, she had symptoms of severe headache, vomiting, and problems with her balance. A CT scan revealed an abscess in the part of the brain responsible for coordination. When the abscess was drained and examined, four types of oral bacteria were found. Even though the woman recovered, it should be remembered that "piercing" creates a "wound" that is subject to infection, especially if the procedure was done improperly, or if proper precautions are not taken while the wound is healing.

250

In the 17th century, people carried torches filled with aromatic herbs in false hopes of using the foul fumes to keep plague at bay. There were not enough coffins for the dead, so lead crosses were placed on victims instead.

251

During the Great Plague of London
in1665, more than 1/3 of London's
half million people died.

252

U.S. outbreaks of plague include: California, 1899
and 1906; Oregon, 1934; Utah, 1936; Nevada,
1937; Idaho, 1940; New Mexico, 1949; Arizona,
1950; Colorado, 1957;
and Wyoming, 1978.

253

Rabies is sometimes called hydrophobia (fear of
water) since an infected person cannot swallow.

254

A 1916 polio epidemic in the northeastern
United States killed 6,000, mostly children, and
left 27,000 with disabilities. Epidemics also
occurred in most years between 1936 and 1954.

255

Vampire bats, rabies carriers in the tropics, live up
to their scary name by sometimes biting humans
as they sleep.

256

In January 2003, a researcher at a Texas Tech University lab was arrested in conjunction with the disappearance of 30 vials of plague bacteria. As it turned out, the bacteria had been destroyed. Plague bacteria is widely available at research labs throughout the world. It is used to study possible vaccines that might be used against it. Plague is considered a bioterrorism threat. Scientists, especially those in the former Soviet Union, have developed methods to mass-produce plague bacteria and turn it into an aerosol. The Soviet Union produced tons of aerosolized plague, including a version that resists the antibiotic tetracycline.

257

In 2002, the civic service organization, Rotary International, received the $1 million Award for Global Health from the Bill and Melinda Gates Foundation for its work since 1985 in eradicating polio worldwide.

258

In the past, one doctor's recommended cure for rabies was to fill an infected person's bite with gunpowder and set it on fire!

259

Progressive Multifocal Leukoencephalopathy
(PML) is a rare infection of the brain by the
JCpolymavirus. This disease most often affects
people, especially men, who have AIDS, leukemia,
or lymphoma. One can have the virus with no
symptoms, until something kicks in (usually
another medical condition) to cause symptoms to
appear. These symptoms, which appear suddenly
or gradually, include paralysis affecting half of the
body, dementia, and possible difficulty speaking, or
blindness. Although some people have survived,
death is common 1-6 months after symptoms
begin. There is no treatment for this disease.

260

Polio outbreaks still strike children in Africa and
Asia; 20 nations are still affected by this disease,
Including Afghanistan. There has been a 99%
decrease in cases since 1988, down to 3,500
cases internationally in 2000. With
continued efforts, complete eradication
is hoped for by 2005. Part of America's
humanitarian efforts in Afghanistan include
polio immunization of children.

261

Once, people were so afraid of polio that children were often restricted from swimming pools and anything else that parents feared might be a source of the disease. If a child was diagnosed with polio, it was not unknown for the parents to drop off the child at a treatment center (such as Warm Springs, Georgia, where U.S. president Franklin Delano Roosevelt was once treated) and never return. While this may seem cruel, parents were driven by the fear that all their children would be infected with the disease.

262

In early 2003, scientists reported a return of polio in India. The reason for a comeback of a disease which was once nearly ended in the country appeared to be false rumors that the vaccine was really a government attempt at population control by causing the inoculated children to grow up sterile. Belief in such rumors caused many parents to refuse vaccines for their children or to fail to complete the entire set of four vaccinations. Most of the reports of new polio cases were in children less than two years old.

263

Q Fever is an infectious disease
(Coxiella burnetii) transmitted by inhaling
infected droplets containing the rickettsiae germ
or by drinking infected raw milk. This infection is
found throughout the world.

264

Two rare cases of "inhaled rabies"
occurred when spelunkers breathed the air
of a bat-infested cave.

265

In 2002, hundreds of kids in at least seven states
broke out in a mysterious red, itchy rash that
seemed to disappear once the students left
school. Although a yet-to-be-identified virus was
suspected, doctors agreed there may never be an
explanation for the cause.

266

Rheumatic Fever is a serious disease that can
affect the valves of the heart. The disease is
uncommon in the U.S. It usually affects children
ages 5-15 who have untreated strep throat.

267

Rat-bite fever mostly affects those who live in slums or ghettos, the homeless, and those who work in biomedical laboratories. Around 10% of rat bites cause rat-bite fever. Bacteria that live in the throat of healthy rats is the most common cause of the disease in the U.S. Weasels and other rodents can also carry the disease. In 1996, a case of rat-bite fever (also known as Haverhill fever) was reported after a teenager drank water from an open irrigation ditch possibly contaminated with rat feces.

268

Following World War II, an epidemic of relapsing fever (spread by body lice or soft ticks) involving about 10 million people, occurred as a result of the massive breakdown in public health systems.

269

In 1989, several cases of relapsing fever were reported after campers spent the night in a cabin built over ground-squirrel burrows at Big Bear Lake in San Bernardino County, California. Other cases had been reported in 1968 in Spokane County, Washington, and in 1973 in Grand Canyon National Park.

270

Dr. Howard Ricketts, first to identify the source of the disease ricketts, died in 1910 in Mexico from another rickettsial disease, typhus.

271

Rocky Mountain Spotted Fever is the most severe and most frequently reported rickettsial illness in the United States. The Rickettsia rickettsii bacterium causes this disease, which is spread to humans by ticks. The disease was first noted in 1896 in the Snake River Valley of Idaho. Today, it has been reported in all U.S. states except Hawaii, Alaska, Vermont, and Maine.
It was originally called "black measles" for its rash. Because the disease was so deadly, the Rocky Mountain Laboratory was established in Hamilton, Montana to study the disease. Today, the disease is treatable with antibiotics.

272

Rotaviruses are the main cause of diarrhea in young children. Virtually all children are infected with a rotavirus before they start school. Scientists suspect that there may be a possible connection between early rotavirus infection and diabetes.

273

Respiratory Syncytial Virus (RSV), like a cold, is an infection of the upper respiratory tract, and, like pneumonia, of the lower respiratory tract. It is the most common cause of pneumonia in children two years old and younger. According to the CDC, almost 100% of kids in childcare get RSV during the first year of life. The infection can range from mild to deadly.

274

In early 2003, it was reported that while medical experts once believed that RSV only affected children, around 11,000 people (78% are 65 and older) die each year from RSV.

275

Rubella, also called German or three-day measles, is close to being wiped out in the U.S., with as few incidences as 272 in 1999. Once, this childhood disease struck fear in the heart of pregnant women because it could cause heart damage, blindness, deafness, or mental retardation in a fetus, as well as a miscarriage or stillbirth if a woman is infected during the first three months of pregnancy.

276

Salmonella, a very common cause of food poisoning, can come from infected eggs, chickens, turtles, lizards, some reptiles, and birds.

277

In 1981, pre-cooked roast beef turned out to be the culprit in a salmonella outbreak in New Jersey. About 100 people became sick. In 1983, a salmonella outbreak in four Midwestern states was traced back to animals on a South Dakota farm.

278

A cult in The Dalles, Oregon sprayed area restaurant salad bars with salmonella in 1984. Their goal? To sicken voters in order to win a local election!

279

SAR11 is a type of bacteria so common in the ocean it is said to be the most common bacteria on earth. SAR11 accounts for 1/3 of all the cells present on the ocean's surface. Just how many cells is that? It is estimated to be 240 x 1 billion billion billion!

280

Scarlet fever can result from an untreated strep throat infection. The disease is contagious and most often occurs in children ages 2-10.
If untreated, scarlet fever can lead to abscessed tonsils or rheumatic fever.
The disease can be cured with antibiotics.

281

Septicemia is blood poisoning. This disease is caused by the bacteria Neisseria meningitidis, which only humans carry, in the mucous of the nose and throat.

282

A 1940 instance of septicemia in a London policeman led to the discovery of antibiotics. A mild scratch became infected and turned into a serious case of blood poisoning. His face, eyes, and scalp swelled and his temperature climbed to 105 degrees Fahrenheit (40.5 degrees Centigrade). Dr. Charles Fletcher gave the dying man doses of penicillin and the patient improved rapidly. However, the drug was only a crude version of the powerful antibiotic we know today, and the doctor was making it one batch at a time. When there was no more penicillin to give him, the man died.

283

Shingles, a more common name for
herpes zoster, is the same as the germ
that causes chickenpox. There is no cure
for shingles. Shingles is most
common after the age of 50.

284

In people with impaired immune
systems, sinusitis can be fatal.

285

The tsetse fly, which causes "Sleeping Sickness,"
is related to the common housefly. It
can carry the trypanosoma parasite in its
salivary glands for 96 days. During this period, it
infects every animal or person it bites.
The flies swarm so fiercely that human
occupation of a place is impossible.

286

In 1977, smallpox became the
only illness ever wiped out as a human disease.
Because the bacteria has no animal host, no
natural outbreak can occur.

287

One of the "gifts" that explorers brought with them to the New World was smallpox. After the Spanish colonized Mexico in the 1500s, within 20 years more than half of the native population had died from this disease. Blankets infected with the smallpox variola virus were sometimes given as gifts to Native Americans in an (successful) effort to kill them off easily and quickly.
In 1775, Boston and Philadelphia endured smallpox outbreaks. Although George Washington had his troops inoculated, the disease continued to make its way west as far as the California coast, and as far north as Canada.

288

One hundred years before the smallpox vaccination was created, Chinese doctors compacted fleas from cows into tablets that became the first recorded example of oral vaccination!

289

Gonorrhea can infect the urethra, cervix, rectum, throat, or whites of the eyes and can spread via the bloodstream to other parts of the body.

290

In 2001, following the terrorist attacks on America, the U.S. government contracted to have 250 million doses of smallpox vaccine made within 12 months. The vaccine can prevent the disease if given within four days of exposure. Amid much controversy, inoculations began in 2002 with military and health care personnel. Doctors were also trained to recognize smallpox, since most would have never seen a case!

291

Smallpox killed more than 500 million people in the 20th century. The last U.S. case occurred in 1949; vaccinations stopped in 1972. Opinions vary on any immunity still existing for those who were inoculated. A diluted form of the vaccine to give some immunity is being tested.

292

In 2002, researchers discovered an experimental drug, Cidofivir, that inactivates smallpox and other viruses in test tubes, could have potential as a future oral vaccine against smallpox used as a bioweapon.

293

Genital warts are caused by the same
virus that can cause cervical cancer.

294

In early 2003, test vaccinations of smallpox were
begun in anticipation of possibly extending the
immunization program. The inoculation
procedure is the same as when it was once
given—the vaccine is "scratched" into the skin,
versus given by injection. A decision for mass
vaccination is controversial since statistics show
that side effects (for .09% or around 250,000
Americans) range from rash to encephalitis; and
for possibly 6 per one million people, death.

295

In early 2003, the U.S. armed forces reported
no deaths from smallpox vaccinations during the
48 years that they were administered
(1942-1990). Therefore, it was concluded that
adverse reactions from smallpox vaccines today
would be much less for people who had already
received the vaccine in the past, with
the most reduced risk the more recently
the inoculation had been given.

296

Who should not get the smallpox vaccine?
In general: anyone with a history of eczema
or other skin infections; people with immune
deficiency such as that caused by HIV, leukemia,
chemotherapy, other medication, or a
hereditary condition; pregnant women; or people
allergic to the vaccine's ingredients.

297

A sneeze is one of the most effective
ways to spread germs! When air passages
become irritated, the body responds with a
sneeze. A sneeze is made up of a fine spray of
mucus and tiny water droplets—packed with
infectious diseases if you are sick. This spray can
be shot out of the mouth and nose at speeds of
up to 100 miles per hour (160 kmph), traveling as
far as 6 feet (2 meters). Because this fine spray is
so easily inhaled by others, teaching children to
cover their mouth and nose with their hand
when they sneeze is a good idea, but using a
tissue is better. When a kid sneezes into his or
her hand and then touches other people or
objects, they only spread germs faster and farther.

298

About 70% of disease outbreaks traced to swimming pools involve the chlorine-resistant organism cryptosporidium.

299

Surgery increases the risk of a "staph" infection. Stitches may become infected. A "post-op" staph infection may progress to toxic shock syndrome if not treated in a timely manner. In 2002, clinical trials of a staph vaccine reduced infection in hospitals by half.

300

Syphilis is caused by the Treponema palladium bacterium. It gets into the body via broken skin or mucus membranes in the genitals. A small, painless ulcer usually develops then heals in 6-8 weeks. Next a rash breaks out, accompanied by headaches, aching bones, loss of appetite, fever, fatigue, and hair loss. If untreated, these symptoms can last for years. Bone and tissue damage, as well as dementia and death, can result.

301

Sporotrichosis is an infection caused by a fungus often found on rosebushes, barberry bushes, sphagnum moss, and other gardening mulch material. As you might imagine, farmers, gardeners, and horticulturists are most often infected. This disease usually affects the skin and lymph vessels nearby, but it can also affect the lungs and other tissues of the body. Symptoms include a small nodule forming on a finger; the nodule grows and forms a sore. The infection then spreads throughout the body. Most of the time, the infection is mild and easily cured with drugs. In rare cases, the disease can affect the spleen, liver, kidney, genitals, or brain.

302

Evidence of tuberculosis has been found in Egyptian mummies and Stone Age skeletons. At one time, some people believed TB was caused by vampires.

303

When immigrants passed through Ellis Island between 1892 and 1954, they were screened for diseases such as typhoid and cholera.

304

Streptococcal infections are caused by various types of bacteria. Group A can cause strep throat, tonsillitis, skin infections, blood infections, scarlet fever, pneumonia, rheumatic fever, St. Vitus' dance (Sydenham's chorea), and inflammation of the kidneys. Group B mostly affects newborns with infections in the joints and heart. Group C and G are often carried by animals, but can infect the human throat, intestine, vagina, and skin. Group D grows normally in the lower digestive tract and the vagina but can cause infections in wounds and heart valves, the bladder, abdomen, and in blood. Some types of strep infections can cause the body to attack its own tissues, leading to rheumatic fever, chorea (jerky body movements), and kidney damage.

305

There have been some reports of tuberculosis being spread on airplanes where air circulation is poor and the flight is long. While this is rare, 1/3 of the world's population is infected with tuberculosis bacteria, some of whom are likely to be among the 1.4 billion air travelers each year.

306

Streptococcus aureus (Strep A) is
a common bacterium. About 1/3 of us
have Strep A germs on our skin. If this germ
wheedles its way into the body through
a cut, during surgery, for example, then blood
poisoning can result. Strep A is a
problem in many hospitals and is resistant to the
powerful antibiotic methicillin. In 1998, more
than 1,000 Americans died of Strep A. More than
10,000 had less virulent forms of the disease.
One St. Louis, Missouri woman survived the
disease, but not before she endured respiratory
and kidney failure, an infected liver, blood
poisoning (sepsis), open wounds from severe
skin sores, shock, the loss of her hair and
nails, and 54 days in the hospital!
She was 38 years old and a healthy non-smoker
when she contracted the infection.

307

Most of the estimated 2,000 carriers of typhoid
in the United States are elderly women with
chronic gallbladder disease.

308

"Just when you thought it was time to get back in the water ..." Pseudomonas bacteria cause two common infections— swimmer's ear and hot-tub folliculitis. These are usually minor infections that affect healthy people. However, pseudomonas can infect the blood, skin, bones, ears, eyes, urinary tract, heart valves and lungs. Burns can become severely infected, leading to a blood infection that can end in death. Serious pseudomonas infections generally occur in the hospital. The bacteria is most often found in moist areas, such as sinks and urinals, although, surprisingly, they can be found in antiseptic solutions!

309

A typhus epidemic killed thousands of Jews incarcerated in concentration camps in Germany during the last days of World War II.

310

A vaccine may be made from a "live virus" or a "killed virus," which cannot infect those getting the vaccine.

311

Germs that resist the immune system
or drugs are becoming known as "superbugs."
These may be new, emerging germs we have not
seen before, or germs that grow resistant to even
the most powerful antibiotics. These superbugs
can produce new diseases. Some of these new
diseases are actually developing in hospitals.
MRSA (Methicillin-Resistant Staphylococcus
Aureus) is one example. This bacterium is found
on the skin and has become resistant to even the
strongest antibiotics. Already weak from illness,
patients are vulnerable to MRSA. Keeping
hospitals spotlessly clean is a priority
in keeping this disease in check.

312

Thrush is a type of yeast infection caused by the
Candida germ. Thrush primarily affects children,
including newborns who get the disease from
their mother, but adults can also get thrush.
Oral thrush is an aggravating infection of the
inside of the mouth and the tongue, which
become coated with creamy white, curd-like
patches. Thrush can be treated with antibiotics.

313

Tapeworms are intestinal infections caused by several types of tapeworms. Beef tapeworm infection is gotten by eating infected raw or undercooked beef. The adult worms that can end up in your intestine can grow up to 15-30 feet (4.5-9 meters) long! Surprisingly, infection usually causes no symptoms. Symptoms that can occur include abdominal pain, diarrhea, and weight loss. Most common in Africa, the Middle East, Eastern Europe, Mexico, and South America, the disease is rare but does also occur in many U.S. states. Pork tapeworm infection is rare in the U.S., most common among immigrants or travelers from high-risk areas of the world. This worm only grows 8-10 feet (2-3 meters) long. Fish tapeworm infections, caused by eating undercooked freshwater fish, occur in many places, including Alaska, the Great Lakes region, and Canada. Treatment comes from various drugs. Actually, there are many more types of worm infections, including threadworms (southern U.S.), pinworms (worldwide, especially in children), dwarf tapeworm (southern U.S.), Echinococcosis (Alaska, Utah, Arizona, Nevada), and flukes (intestinal, sheep liver, fish, lung, and blood). Prevention? COOK! COOK! COOK! Occasionally, a person with a tapeworm may feel a piece of the worm wiggle out through the anus! Hello!!

314

Tetanus is caused by the Clostridium tetani bacterium. This disease is also nicknamed "lockjaw." Because of immunizations, it is rare in the United States. The disease is difficult to treat. It is not contagious. These germs are common in dirt, but die quickly when exposed to oxygen. This is why any cut contaminated with soil should heal in open air. An unimmunized person usually gets tetanus after stepping on a dirty nail or being cut by a dirty tool. The bacteria infect the wound and produce a toxin that spreads through the blood. This toxin can cause severe muscle spasms, paralysis, and often, death. Adults should get a tetanus booster every ten years.

315

In January of 2002, at least 48 people connected to Spring Hill College in Alabama tested positive for tuberculosis after a student from Kenya died of the disease the previous December. Earlier in the month, five immigrants in Mecklenburg County, North Carolina were diagnosed with a drug-resistant form of tuberculosis.

316

A tick is a blood-sucking parasite. Mature ticks usually live and feed on deer. However, they also bite humans. A tick will feed 2-4 days before dropping off. This is why it is imperative to examine yourself, and especially children, for a possible tick bite after being outdoors in the woods. Pay special attention to the scalp hidden beneath the hair. The Ixodes tick is the source of Lyme disease, an arthritis-like illness. Ticks can also be the source of babesiosis, relapsing fever, rickettsial diseases, and Rocky Mountain spotted fever.

317

Doctors believe that getting up to 20 vaccinations by age two does not increase a child's risk of a weakened immune system.

318

Although there is not cure for genital herpes, recent tests of a vaccine reduced the risk of infection by 75%. However, the vaccine is effective only in women. This is the first known instance of a vaccine working in one sex, but not the other.

319

Toxic shock syndrome is an infection usually caused by the staphylococci bacteria. If not quickly diagnosed and treated, it can result in shock and death. In 1978, toxic shock syndrome was first recognized as something that could affect children ages 8-17. However, by 1980, it was apparent that young women—almost all of whom were tampon users—were most commonly affected by the syndrome. After as many as 700 cases were reported in just a little over a year in the United States, "superabsorbent" tampons were removed from the market. As a result, toxic shock syndrome cases dropped dramatically. Cases still occur in women who do not use tampons, in women who have just had surgery or given birth, and in men who have had surgery. Between 8-15% of those affected by toxic shock syndrome die. It is unclear what causes toxic shock syndrome to develop. A tampon may increase the likelihood that bacteria will produce a toxin that enters the blood through small cuts in the vaginal lining or through the uterus.

320

Toxins are poisons produced by living organisms; there are also synthetic, man-made toxins. These toxins can be gaseous, liquid, or solid. When toxins are used for military purposes, they are classified as "chemical warfare agents." Today, thousands of poisonous chemicals exist, but only a few are actually considered suitable for use as weapons of war. Furthermore, while 70 different chemicals have been used or stockpiled as "CW" agents during the 20th century, only a few are considered serious as potential warfare agents today. To be usable as a weapon of war, a toxin must also be able to be handled, stored, and resistant to water, oxygen and heat. Depending on the toxin, humans can be harmed by inhaling an agent, contact with the skin, or contaminated water or food. Because these toxins affect the nervous system, they are often referred to as nerve agents. Some of these nerve agents include: Tabun, Sarin, Soman, and VX. (I would give you their full names, but mostly they are an unpronounceable alphabet soup!) Reaction to exposure to nerve agents depends upon the strength of the agent. Symptoms can range from increased saliva production, runny nose, chest pressure, and contracted pupils ... to convulsions and death.

321

Toxocariasis is an infection caused by the roundworm Toxocara canis or Toxocara cati. These parasites' eggs can be found in sandboxes, where cats defecate. Children can become infected with eggs contained in the feces. After being swallowed, the eggs hatch in the intestine and the larvae are spread throughout the body via the bloodstream. Inflammation of the brain, eye, liver, lung or heart can result. In young children (ages 2-4), symptoms may be mild; older children and adults can also become infected. Symptoms include fever, cough, wheezing, liver or spleen enlargement, skin rash, and pneumonia. Eye lesions may occur in older children. Treatment is with antibiotics, although most infections go away on their own in 6-18 months.

322

President Thomas Jefferson was one of the first advocates of vaccination.

323

Germs found in wells and public water sources can include parasites such as giardia and cryptosporidium, as well as bacteria such as E. coli and salmonella.

324

Toxoplasmosis is an infection caused by a single-celled parasite. These parasites are found in the intestines of cats, and people may be infected after coming into contact with cat feces. Infection can also come from eating contaminated raw or undercooked meat. This disease is especially dangerous for pregnant women and their unborn children. There are several forms of the disease, including mild lymphatic toxoplasmosis, chronic toxoplasmosis, and acute disseminated toxoplasmosis, which mostly affects people with an impaired immune system. People with AIDS can get a more serious version of the disease. Treatment is generally with antibiotics.

325

In early 2002, unprecedented shortages of even ordinary childhood disease vaccines were reported, forcing some doctors to postpone some regular immunizations until vaccines became available.

326

Nearly 75% of women will endure at least one vaginal "yeast infection" during their lifetime.

327

Tropical Spastic Paraparesis (TSP) is a
viral infection of the spinal cord. It causes
weakness in the legs. The virus that causes this
disease can also cause a type of leukemia. TSP
can be spread by sexual contact or infected
needles, and from mother to child via the
placenta or breast milk. Symptoms may not begin
until years after the infection. Symptoms include
muscle weakness and stiffness in both legs;
sensation in the feet may be lost. Although
there is no cure, people treated with
corticosteroids may improve quite a bit.

328

You may remember that French chemist Louis
Pasteur discovered that animals vaccinated against
a disease were protected from catching that
disease, but did you remember that his
experiment was done with anthrax? Pasteur also
developed a rabies vaccine, with which he saved
the life of a boy bitten by a rabid dog.

329

Scientists are investigating the possibility that
viruses play a role in obesity.

330

Trichinosis is the thing my mother used to scare me with so that I learned to only eat "crispy" (meaning nearly burned!) bacon. This disease is an infection caused by eating raw or undercooked pork or pork products, bear, boar, and some marine mammals. It is caused by a parasite. This parasite eventually finds its way into certain muscles, such as the tongue, eye, between the ribs, and the heart. Symptoms include swelling of the upper eyelids, muscle pain, rash, and bleeding beneath the fingernails. Difficulty swallowing may follow, sometimes even causing death. Most people fully recover. Prevention is best achieved by eating only thoroughly cooked meat. Trichinosis occurs in most parts of the world. It is now rare in the U.S., and is rare or even absent in places (such as France) where pigs are fed root vegetables.

331

Antibiotics do not work against viruses. Vaccines do exist to prevent some viral diseases. Scientists are working on antivirals, which are not as powerful as antibiotics, but may offer some defense against viruses.

332

Trichuriasis is an infection caused by an intestinal roundworm. This disease mostly occurs in the tropics and subtropics. People are infected by eating food contaminated with roundworm eggs, which hatch in the small intestine then migrate to the large intestine where they burrow their heads in the intestinal lining. Each larvae grows to about 4.5 inches (11.4 centimeters) long. Mature female roundworms produce around 5,000 eggs each day, which are passed in bowel movements. Infection can cause symptoms as mild as abdominal pain and diarrhea to more serious problems such as bleeding from the intestine, anemia, weight loss, and appendicitis. No treatment is given for mild infections; drug treatment is given for more serious infections. Good personal hygiene, sanitary toilet facilities, and avoiding unclean vegetables are good prevention methods.

333

Some Victorian women intentionally infected themselves with tapeworms to develop malnutrition, so that they could maintain the tiny waistlines popular during the era!

334

Tuberculosis (TB) is caused by the Mycobacterium tuberculosis bacteria and spreads when an infected person coughs, sneezes, or talks and releases tiny droplets of microbe-filled water that is inhaled by others. TB is common in crowded cities and filthy conditions. TB was reduced following the development of pasteurization, which effectively killed tuberculosis bacteria in cows' milk. In 1922, streptomycin became the first TB vaccine. However, due to antibiotic resistance, tuberculosis has become more difficult to treat. Sources of TB include infected people, poorly ventilated, unsanitary places, and unpasteurized milk. Cattle, pigs, birds, and badgers sometimes carry TB bacteria. TB spreads through the air when an infected person coughs, sneezes, yells, or sings. It is not spread by objects the infected person has touched. A child may get TB from a person in the home who has active TB. A person with TB infection, but not active TB disease, is not contagious. Symptoms include a painful, racking cough, weight loss, and paleness of skin color. "TB infection" means you have the germ in your body even though you do not have any symptoms. "Active TB" or "TB disease" is when you have the germ and symptoms. Diagnosis is with a TB skin test, a safe, quick test that can be given at your local health department. It's a simple pinprick on the forearm and results are available in 1-2 days.

335

In 2001, an estimated two billion people were believed to be infected with TB worldwide, with eight million new cases each year, and two million TB deaths annually, making tuberculosis one of the most common infectious diseases globally. Multi-drug resistant TB has been noted in 35 countries. In the United States, TB is more common among immigrants from Asia, Africa, and Latin America, as well as minority groups that may not receive timely or adequate medical care. Some states require children to have a TB test before they are allowed to start school.

336

It is believed that one way to reduce the risks of food poisoning is to drink wine with a meal. According to scientists "wine kills bugs" including food-borne disease-causing germs such as E. coli, salmonella, Staphylococcus aureus, Psuedomonas aeruginosa, and Klebsiella pneumoniae.

337

The word virus is Latin for "slimy liquid" or "poison."

338

Tularemia is one of the most infectious diseases known. In the 1950s and 1960s, the United States studied tularemia as an infectious organism suitable for use as a bioweapon. This germ infects humans by entering through the skin, the mucus membrane of the eye, the lungs, or the gut. There are several types of the disease: ulceroglandular (ulcers on the hands, fingers, and lymph nodes); oculoglandular (infecting the eye and lymph nodes; glandular (lymph nodes); and typhoidal (often leading to pneumonia). Without antibiotics, 1/3 die. A vaccine is under review by the U.S. Food and Drug Administration. Francisella tularensis was first discovered in Tulare County, California in 1911. It is also referred to as rabbit fever or deer fly fever.

339

Yaws is a disease that is very similar to syphilis, but it is not sexually transmitted. It is caused by the bacterium Treponema pertenue. Yaws is generally found in tropical and subtropical regions near the equator. It almost always occurs in poor children.

340

Before people learned that filth was a source of infectious disease, few bothered with personal hygiene. Water infected with human waste was, and still is, one source of outbreaks of the disease typhoid. The Industrial Revolution of the 1700s and 1800s helped create many large, crowded, unsanitary cities. In the 19th century, typhoid outbreaks killed hundreds each year in New York City and London. Between 1900-1915, New York City cook Mary Mallon infected at least 53 people with typhoid before being forced into an isolation hospital. "Typhoid Mary" was an example of a "carrier"—someone infected with a disease who passes it on, but has never shown symptoms him- or herself.

341

In the early 1900s, work on the Panama Canal had to be halted because of so many workers dying of Yellow Fever, until it was discovered that mosquitoes breeding in the canal were causing the disease.

342

Urinary tract infections can be caused by bacteria, viruses, fungi, or parasites. The most common way such infections enter the urinary tract is through the opening at the tip of a man's penis or the opening of a woman's urethra. Infection can also come directly through the bloodstream. Bacterial infections of the bladder and urethra are very common, especially in women. Most of these infections are caused by bacteria from the person's own intestine or vagina.

343

In 2003, the World Health Organization (WHO) reported that the lives of millions of children could be saved if hygiene, sanitation, and the availability of vaccines were improved in developing countries. The poorest 20% of the world's population, in places such as sub-Saharan Africa, account for 50% of child deaths from diseases such as whooping cough, polio, diphtheria, measles, and tetanus.

344

Almost every person produces stomach acid. However, only about one in ten people will develop peptic ulcers. An ulcer develops when the normal defenses that protect the stomach or duodenum (the first few inches of the small intestine just below the stomach) from stomach acid break down fail. A peptic ulcer is a roundish sore created when the lining of the stomach or duodenum is eaten away by stomach acid. Today, we know that stomach ulcers are not caused by acids, but by bacteria. Helicobacter pylori appears to be the cause of many stomach ulcers. It is unknown how these germs create the ulcer—whether they disturb the protection against stomach acid, or perhaps produce a toxin that contributes to the formation of an ulcer.

Such ulcers are now generally treated with antibiotics instead of antacids. Researchers have found that sulforaphane, found in broccoli and broccoli sprouts, kills H. pylori bacteria. This natural compound apparently is effective even against antibiotic-resistant forms of the bacteria.

345

Viruses are the smallest of the microbes. They can cause flu, polio, Yellow Fever, smallpox, and other diseases. To survive, viruses have to take over other living cells. This is why they are such a problem—the cells they have invaded must be killed in order to kill the virus. A virus is a strand of genetic matter covered with a protein shell. The genetic material inside a virus lets it copy itself once it has invaded a cell. A virus may also be covered with proteins called antigens that help it invade cells. Our immune system recognizes a virus by its antigens. When a virus reproduces, it often changes (its antigens, too), so the virus may go undetected in our body. A large change in the antigens of a virus may lead to an epidemic because so few people would have any immunity to the new virus. This is why there seems to be a new flu vaccine each year—the flu virus is good at changing easily and quickly, creating a new "strain" of flu.

346

In 1793, Yellow Fever killed 5,000 people in Philadelphia; 2,000-3,000 died of the disease in New York City in 1798.

347

Well, I'm sure most of us never thought we'd have to be afraid of our mail (bills and credit card solicitations excepted), and now ... But do you really have to be afraid of opening your mail? The odds are, that unless you are a news anchor or politician, probably not. Just in case, here are warning signs the U.S. Postal Service says you should watch for:

- No return address
- Postmarked from a foreign country
- Misspelled words, poor typing or handwriting
- Excessive postage
- Rigid or bulky
- Excessive tape or string
- Odd odor
- Wrong name or title
- Oil stains, discolorations
- Protruding wires (no joke!)
- Restrictive markings such as Personal, Special Delivery, Confidential
- Powdery exterior

And what if you do get one of these types of letters or packages? Put it down and leave the area, call a postal supervisor, call the police, wash your hands, remove your clothes and place them

(347) in a sealed plastic bag, take a shower, make a list of anyone else who was in the area when you handled the letter or package. The odds are your poorly handwritten letter that stinks to high heaven is from Uncle Ed who addressed it on the work table in his garage. Be realistic, be calm. Mail is still mostly mail.

The CDC has no comment that ironing or microwaving your mail offers any proven decontamination effect. Translation: don't bother? However, the U.S. Postal Service has begun disinfecting some mail to kill any biological agents. This "irradiation" is the same technology already in use to sanitize food and medical equipment. Irradiation makes food safer from bacteria that can cause food-borne illness. Irradiation is endorsed by the World Health Organization, the CDC, and the American Dietetic Association.

348

BioPort, in Lansing, Michigan, is the only U.S. facility producing the anthrax vaccine, all of which is reserved for the military at this time.

349

Warts, also known as verrucae, are caused by the papillomavirus virus. Warts can spread from one part of the body to another, but are not very contagious from one person to the other (with the exception of genital warts). Most warts are a nuisance and not harmful; they generally disappear over time without treatment. Even with treatment, warts may return. Warts are most common in children, and less common in the elderly. Only rarely are warts cancerous. Warts that grow in clusters are called "mosaic" warts. Almost everyone gets common warts, known medically by the unattractive term verrucae vulgaris.

350

Chronic wasting disease in deer and elk has been found in Montana, South Dakota, Nebraska, Colorado, Kansas, Oklahoma, New Mexico, Illinois, Wisconsin, and Alberta and Saskatchewan, Canada. Usually months to years go by before infected animals show signs of the disease. Some states have elected to kill off infected deer to slow the spread of the disease. No one should intentionally eat deer or elk meat known to have come from infected animals.

351

In the 1990s, three outdoorsmen
died as a result of a disease which destroyed their
brains. A disease known as "chronic wasting
disease" had previously affected wild game in the
area. The men had eaten elk and deer meat in
Wisconsin in the 1980s and 1990s. Although
there has never been a documented case of a
person contracting a brain-destroying disease
from eating wild animals infected with chronic
wasting disease, public health investigators began
research in 2003 to learn whether or not such a
disease had crossed from animals to humans.
Such a thing occurred in Europe when "mad cow"
(new variant Creutzfeld-Jakob disease)
disease caused brain-destroying illnesses in
humans, many of whom later died.

352

The Yellow Fever virus was once found only in
animals. The first instance of humans contracting
this virus was recorded in 1684. Yellow Fever
came to America on board the cramped,
crowded, disease-ridden slave ships from Africa.
Yellow Fever can be mild or fatal. Only about one
million doses of vaccines are available in the U.S.

353

In November 2002, the CDC confirmed that outbreaks of disease from drinking water and swimming pools had risen dramatically in recent years. Reports of outbreaks of contaminated water have doubled in the last few years. For the most recent years statistics were available (1999 and 2000), 39 outbreaks involving drinking water in 25 states were reported. This was in spite of the fact that improvements had been made in publicly operated water systems. The CDC warned that whether from the tap or a bottle, people should be concerned enough about their water supply to know where it comes from and how safe it is.

354

The U.S. Army Medical Research Institute of Infectious Diseases, located at Fort Detrick in Frederick, Maryland, is also called USAMRIID or "the Institute." They call their morgue "the Submarine" and their hot zone hospital "the Slammer." I guess it helps to have a sense of humor when dealing with deadly stuff all day?

355

Worms are some of the largest creatures that can cause disease in humans. These parasites can grow as long as 40 feet (12 meters). The worms that infect humans are either flatworms or roundworms. People become infected through worm eggs, larvae, or from infected food, water, or soil.

Types of worms:

- **Common roundworm:** Pencil-size; lives in intestines; children get from infected dog or cat feces (or soil tainted with them)

- **Bilharzia flatworm or fluke:** Schistosomiasis disease enters the skin from infected water; travels through the blood vessels to major organs where they lodge and mature; symptoms include fever and diarrhea; there is no vaccine; severe damage to the body can result

- **Tapeworm:** Attaches itself to the intestinal wall; infection comes from infected, undercooked meat

- **Hookworm:** Small, bloodsucking roundworms that infect the wall of the small intestine; gotten via larvae through the skin

356

While most people associate yeast infections with women's vaginas, such infections can also affect the male penis, cause the mouth infection known as thrush, cause diaper rash, and infect the bloodstream. Yeast infections are caused by various germs that are commonly found on the body, but are doing no harm. However, certain changes can upset the balance of these microbes and create infections. Such changes include the use of antibiotics or excessive moisture. In the elderly, certain antibiotics or steroids inhaled for asthma treatment may bring on yeast infections.

357

Zoonotic diseases, or zoonoses, are those that can be transmitted from animals to humans. Even animals that appear healthy may carry infectious diseases. These diseases can be transmitted through contact with the animal, its carcass, byproducts, or contaminated habitat. Zoonotic diseases are caused by bacteria, viruses, chlamydia, fungi, and parasites. Animals which can transmit zoonoses include cattle, sheep, horses, pigs, chickens, turkeys, dogs, cats, rodents, and some wild animals.

358

Gangster Al Capone died of an advanced case of syphilis in spite of treatment from doctors at the prison on Alcatraz Island off the coast of California.

359

The National Pharmaceutical Stockpile is a cache of drugs and medical supplies located at eight locations around the U.S. The stockpile is divided into 50-ton (45 metric tons) "push packages" that can be dispatched immediately in an emergency. Each push package can treat 10,000-35,000 people.

360

Cholera outbreaks (in addition to a tainted-milk scandal) in the 1850s, spurred Central Park designers in New York City to build a dairy so children could get fresh, regulated milk. Today, "The Dairy" is used as a visitors' center.

361

Scientists have created an algae that makes antibodies that target the herpes virus.

362

Don't tell the tooth fairy: Scientists believe that stem cells found in baby teeth may be used to treat diseases.

363

In 2003, the U.S. government installed high-tech sensors in large cities that could detect smallpox, anthrax, and other deadly germs during a bioterrorism attack. Such a system was tested during the 2002 Winter Olympics in Salt Lake City, Utah. The CDC also has a system that consists of monitoring visits to doctors offices and emergency rooms, as well as drugstore sales.

364

An October 2001 poll reported confidence in the government to respond to the following health threats:

Outbreak of...

	Very Confident	Somewhat Confident	Not Too Confident	Not at All Confident
Anthrax	34%	43%	17%	5%
Smallpox	27%	39%	2%	10%

365

Germs are important enough to have treaties written to try to keep them under control. Just a few include the following short list. Judge their effectiveness to-date yourself!

1899 Hague Convention banned poisons as weapons of war.

1925 Geneva Protocol banned the use of (but not production of) chemical weapons in war.

1972 Biological and Toxin Weapons Convention, ratified by 142 nations, including the United States; banned the production, possession or use of germ warfare agents.

1977 Chemical Weapons Convention said all nations must destroy their stockpiles of chemical weapons by 2007.

2001 U.S. senators Ted Kennedy and Bill Frist drafted a bioterrorism proposal providing immunity for drugmakers working on bioterrorism treatments and a streamlining of the drug-approval process.

366

In 2002, 1,000 horses in Minnesota were infected with the West Nile Virus. At least 300 died from the disease.

367

A molecule called ubiquitin appears to help control everything from HIV and Ebola viruses to cancer and heart disease. Scientists are just now exploring how this amazing molecule might be used to fight germs.

368

Save money: The CDC says that the best disinfecting solution for toys, diaper changing tables, and other surfaces is a mixture of 10% bleach to 90% water.

369

According to the CDC, for every one million smallpox vaccinations, there will be 1-2 deaths, 15-52 life-threatening reactions, and up to 1,000 milder reactions that can have serious consequences.

370

In early 2003, an emerging epidemic of drug-resistant staph infections broke out among gay men in Los Angeles County, California. One infectious-disease specialist said, "It seems to be able to attack normal skin in healthy people."

371

U.S. President Franklin Delano Roosevelt contracted polio as an adult and effectively hid his paralysis by being photographed and filmed with desks that hid the wheelchair in which he sat. In an effort to find relief from the disease, he visited and even established a "Little White House" in Warm Springs, Georgia where he could take advantage of the area's medicinal waters.

372

Microbes multiply rapidly and constantly. Their population sizes are expressed in numbers that end with 12, 13, or 14 zeroes, for example. The time between generations is only minutes!

373

In 1866, Kamehameha V isolated lepers on a remote part of the Hawaiian island of Molokai. Leprosy was believed to have been brought to the islands by Chinese immigrants. The lepers were forced to leave their homes and travel by ship to the Kalawao peninsula where they were tossed overboard into the waves, either to drown or swim to shore to fend for themselves. In 1873, Belgian priest Father Joseph Damien de Veuster came to live with and aid the lepers. He contracted the disease and died there at the age of 49.

374

The vaccine Prevnar reduces blood infections, meningitis, and some pneumonia and earaches in young children. Seniors are also protected from these diseases when such infections are eliminated or prevented in children.

375

In the spring of 2003, 82 cases of "fowl plague" appeared in the Netherlands, which doctors feared could spread into an epidemic.

376

Some microbes, when stressed, can mutate and behave in other ways that help them survive in spite of harsh environments. Some microbes can now even tolerate pure bleach!

377

Microbes make up about
1/20th of a human's weight.

378

One of every two people in
low-income countries dies at at early age
from an infectious disease.

379

In 1997, the number of cases of tuberculosis in New York City was 55% lower than in 1992. NYC contributed 31% to the national decrease in tuberculosis during that period.

380

Germs kill 17 million people each year.
A third of these deaths occur in
developed countries. This is a 50%
increase since 1980.

381

During the era of smallpox in the 1800s, 40% of infected adults and 90% of infected children died. Those who survived were usually blind for life.

382

Viruses are not truly living organisms. The only way they can operate is to gain entry into a cell before they can move or reproduce.

383

Of the 100,000 species of fungi, more than 400 are known to infect people, causing everything from inconsequential irritation to death.

384

The CDC maintains a division of vector borne infectious diseases in Fort Collins, Colorado.

385

Aflatoxins come from fungi that grow on peanuts. As the fungi grow, they release a toxin which can cause liver damage when contaminated peanuts are eaten.

386

How small are viruses? The largest are about 450 nanometers (about 0.000014 inches; 0.0000355 cm) and the smallest are around 20 nanometers (0.0000008 inches; 0.000002 cm). Even using the most powerful electron microscopes, only the largest viruses can be seen.

387

In 1994, a 1,000% increase in infant deaths from an unknown lung disease alerted doctors to a problem in an area east of Cleveland, Ohio. Most of the victims were boys under six months old. The culprit germ was eventually determined to be Stachybotrys chartarum, a fungus that thrives in damp wallpaper and wet cellulose-containing building materials.

388

Why don't antibiotics work against viruses? Because antibiotics work by taking out some of the biochemical processes inside bacteria. Viruses are so primitive that they don't have any processes to take out!

389

In the 1700s in Berlin, Germany,
98% of children younger than age five who
contracted smallpox died.

390

What gives bubonic plague its name?
Buboes are the large, swollen lymph nodes in the
neck, armpits, or groin that this infection causes. A
pus-covered ulcer can develop at the site of infected
flea bites. Infection of the blood can cause bleeding
beneath the skin which causes the black spots that
gave rise to the "Black Plague" nickname for the
disease. Victims often stretch and flex their arms to
try to relieve the pain radiating from the buboes. As
if these symptoms weren't bad enough, add extreme
fever, headache, shaking chills, and delirium. When
untreated, death occurs in 50-60% of cases.

391

Bubonic plague bacterium can infect a
person's lungs, leading to pneumonic plague.
This disease is highly-contagious because it
can be spread through the air. A severe and bloody
cough can lead to coma and death in almost all
cases if treatment is not begun
within hours of the first symptoms.

392

Septicemia plague occurs when large quantities of Yersenia pestis germs get into the bloodstream. This can bring on rashes, gangrene, breathing difficulties, blood clotting problems, multiple organ failure, and death within a day in this virtually 100% fatal disease.

393

Instead of dying out, plague has rallied with increased incidences in the mid-1960s, 1973-1978, and the mid-1980s. At least 247 cases of plague occurred between 1980 and 1997, with 37 deaths. This is the highest of any 18-year period since the early 1900s. In 1997, there were 4,370 cases reported worldwide, and the World Health Organization (WHO) believes that these numbers are probably much lower than actual incidences of the disease.

394

At various times in history, syphilis was called the Venetian Disease, the French Disease, and the Naples Disease—presumably depending on who you wanted to blame for the scourge?

395

Which famous ancient personages died of smallpox? Pericles (430 B.C.); Ramses V. (1157 B.C.); Hittite King Suppiluliumas I and his heir Arnuwandas (mid-1300s); Spanish prince Baltasar Carlos (1646); William II of Orange and his wife Henrietta (1650); Austrian Emperor Ferdinand IV (1654); Emperor Gokomyo of Japan (1654); Emperor Fu-lin of China (1661); England's Queen Mary II (1694); Ethiopia's King Nagassi (1700); Tsar Peter II of Russia (1730); Ulrika Eleanora, Queen of Sweden (1741).

396

Early attempts at vaccination for variola (smallpox) included spreading infected pox scabs (or pus from such scabs) on the skin or blowing powdered scabs into the nostrils. An Englishwoman, Lady Mary Wortley Montague (who had her beautiful face disfigured by smallpox and lost her 20-year-old brother to the disease), was instrumental in encouraging a vaccination technique she had observed elderly women in Istanbul use, which involved scraping smallpox pus into the skin. To this day, a similar process of multiple sticks in the skin (with killed versus live virus) is used as immunization against this disease.

397

Bacteria can not only pass mutations allowing resistance to antibiotics to subsequent generations ... but can also pass this knowledge to unrelated bacteria who just happen to be living next door to them!

398

In 1978 (the year after the last natural case of smallpox was recorded), an English medical photographer, Janet Parker, died of the disease after contracting it in her office, which was one floor above where smallpox was being stored at the university where she worked. After it was determined that the virus was transmitted through a faulty ventilation system (and that researchers often left their gowns on after leaving their laboratory), around 300 people had to be quarantined. Only the woman's mother eventually contracted the disease and she survived. The head of the department involved was so distraught that he committed suicide.

399

Researchers believe that a virus called Herpes 6 might be a factor in multiple sclerosis.

400

The placenta normally forms an impenetrable barrier to germs. One exception is Listeria monocytogenes which can cause an infection that can kill a fetus. This is why pregnant women are advised not to eat unpasteurized food.

401

In many countries, antibiotics can be bought "over the counter" instead of by prescription-only. This has increased the problems with antibiotic resistance by germs.

402

Taking one antibiotic can trigger multidrug resistance. Tests have shown that bacteria being attacked by antibiotics cannot only develop a resistance to that antibiotic, but to others as well. This included synthetic antibiotics as well as natural ones.

403

Where do viruses come from? Speculation includes everything from bacteria that evolved into simple viruses ... to from meteors from outer space!

404

Scientists are attempting to genetically engineer a plant that will change color rapidly if it senses a biological or chemical weapon in the area. It is anticipated that such plants placed in shopping malls and other public areas might serve as a biohazard early warning system.

405

In late summer 2003, U.S. troops in Iraq came down with pneumonia, possibly as a result of dehydration and inhaling so much dust.

406

From 1770-1892, people on the east coast of North America were so convinced that vampires were the source of TB that they dug up suspected vampires. If blood found in the heart or the body seemed to be decaying too slowly, this "proof" was enough to require the removal of the body's head and its placement on the upper chest. The body's heart was also removed, then burned.

407

Scientists say it is possible that some common disinfectants, including those in toothpastes, deodorants, and soaps, actually act like antibiotics and therefore add to the problem of promoting drug resistance in germs.

408

While some scientists insist that about 1/3 of the prescriptions written each year are actually unnecessary (either because it would have no affect on a virus or because a healthy person's immune system would have fought off the disease on its own), other scientists assess that possibly as few as 1% of such prescriptions are truly needed and necessary.

409

Fortunately, diseased tissue was collected from 78 soldiers who died of flu during the 1918-20 pandemic. Today, scientists at the Armed Forces Institute of Pathology in Washington, D.C. are analyzing these important specimens to be better prepared for the next major pandemic which most scientists believe is overdue.

410

The 1918 Spanish influenza pandemic began with a simple case reported on March 4 at Camp Fuston, Kansas. On March 11, an army private at Fort Riley, Kansas came down with the flu. In one week, the number of healthy young soldiers who became ill in the camp rose from 100 to 500, with 48 deaths. This began the first "mild" wave of the disease which then traveled the world, eventually disappearing in 1920 without a trace, but not before killing millions.

411

During the Spanish influenza epidemic of 1918-20, entire communities in Alaska were wiped out. One family had three teenage sons die in one night. Teller (now Brevig) Mission lost 85% of its residents. People were ordered to wear masks in public and were fined or jailed if they did not do so. One village posted armed guards ordered to shoot anyone who approached the village; this village remained flu free. Some claimed that these deaths were not caused by the flu, but by noxious night gasses, nudity, fish, the Germans, the Chinese, open windows, closed windows, or dirt.

412

Influenza comes in three classes—A, B, C. Think of it like this: C is common and not particularly lethal; B is bad, usually causing more typical epidemics among susceptible people such as the elderly; and A is awful and the source of deadly worldwide pandemics.

413

For a long time, scientists thought that the Ebola virus was transmitted only from animal to human. However, in 1989, infected monkeys in a lab in Reston, Virginia came down sick with Ebola. Scientists were shocked to find that Ebola Reston, as it came to be known, could be transmitted through the air. Fortunately, the disease could only kill the apes.

414

Scientists have shown that bacteria can sometimes commit suicide. They may do this after they are attacked by bacteriophages (viruses that can kill bacteria). Researchers suspect bacteria do that to keep the phages from multiplying and killing all the bacteria in a colony.

415

Dengue fever was only known in 9 countries before 1970. By 1995, epidemics had been documented in 40 nations. By 1998, dengue was endemic in 100 countries, with 50 million cases recorded each year. Caused by a bite from an infected Aedes aegypti mosquito, dengue starts with flulike symptoms. While not all cases advance into dengue hemorrhagic fever, those that do cause high temperature and an enlarged liver. There is no vaccine. Unless the symptoms slowly fade away, victims go into a state of shock and die in 12-24 hours.

416

In February 2002, President George W. Bush began pushing a 10-year $6 billion initiative called "Project BioShield" which would give the government more power to stockpile vaccines, without waiting for congressional approval. It would also allow the federal government to skip the Food and Drug Administration (FDA) certification process for a drug or vaccine in an emergency situation.

417

Roman, Persian, and Greek soldiers were said to have thrown dead animals into the water wells of their enemies to poison them.

418

In Poland, doctors have treated as many as 550 patients (518 with infections resistant to antibiotics) with bacteriophages, achieving good results in 9 out of 10.

419

West Nile Virus symptoms include high fever, sore throat, backache, muscle pain, joint pain, eyestrain, blurred vision, and a bodywide rash; also, anorexia, nausea, vomiting, diarrhea, and breathing difficulties. One in six people can go into a coma and die. In February 2002, It was reported that an eye exam could help determine the presence of West Nile virus in a person.

420

What are the most common vectors, or carriers, of drug resistant germs? Doctors and nurses.

421

The viral hemorrhagic disease known as Marburg is named after a German industrial town. In 1967, three men working at a plant where vaccines were made came down with flulike symptoms. The scientists had been working with vaccines grown on cells cultured from monkey kidneys. Soon the men exhibited symptoms including bright red eyes and swollen spleens. Shortly, 23 additional workers grew ill, as did six in Frankfurt, and a vet and his wife in Belgrade, Yugoslavia. Two other wives came down with the disease. As the patients' illness grew worse, their disease-fighting white cells dropped drastically in number. Their skin, covered in rashes, was so sensitive to touch that they could not even stand a sheet to cover them. All had throats so sore that they could not swallow and had to be fed intravenously. Next, they began vomiting and their skin peeled off, even from their genitals. Some died; one became psychotic; and those that lived suffered chronic liver disease for the rest of their lives. All had had some connection with infected monkeys or those who worked with them. Of the 99 animals in the batch that were worked on, 49 died. Tests showed that the disease had been in the monkeys since as far back as 1961. In 1975, an Australian tourist in southern Africa died of Marburg and his girlfriend and a nurse contracted the disease after caring for him; they survived. In 1999, 70 died of Marburg in the Democratic Republic of Congo; the next year, around 20 people died in an outbreak.

422

In the 1950s, a new disease emerged in the village of Lassa in the Yedseram River valley of northern Nigeria. In 1969, the hemorrhagic disease was diagnosed as being caused by a virus and was named Lassa Fever. The virus originates in the Mastomys natalensis rat which lives in sub-Saharan Africa. You can become infected by breathing in the dried urine of an infected rat. During epidemics, the virus is passed from person to person by direct contact with infected blood, throat secretions, or urine. Because the virus is especially high in semen, the disease can be contracted via sexual intercourse up to three months after recovery. The disease culminates with a massive fever and the coughing up of dissolved lungs and other body parts. Although there is no vaccine, if the disease is caught early it may be successfully treated with the antiviral drug Ribavirin. Often, such treatment is not started in time because Marburg may be diagnosed as malaria in the beginning. Around 20,000-40,000 people become infected each year with several thousand deaths. At least four cases were documented in Europe in 2000 after people who were infected traveled there from Africa. Epidemiologists were kept busy trying to track down anyone who had come into contact with these infected patients.

423

Prions have been defined as infectious lumps of protein. In other words, they are not bacteria, not viruses, but are germs—pathogens that can spread disease. Prion proteins are the culprits behind diseases referred to as spongiform encephalopathies; these include "mad cow," scrapie, and Kuru. Formerly, such brain-destroying diseases were mysterious to scientists. Only now are they learning more about these neurodegerative diseases that cause holes to develop in the brain.

424

At one time, researchers calculated that a maximum number of people affected by New Variant Creutzfeld-Jacob disease will be no more than 166,000. Is this supposed to make us feel good? They say it's better than the millions they originally anticipated!

425

In 1979, at least 62 Russians died following an outbreak of pulmonary anthrax in the Ural Mountains, supposedly after a breach in the weapons research factory located at Sverdlosk.

426

Scrapie gets its name from the fact that animals (sheep, for example) with the disease tend to rub against fenceposts (for example) so hard and so constantly that they scrape off all their wool. The disease is most common in sheep and goats, but mink, elk, mule deer, cats, and cows all have their own versions of the disease. Humans can also contract similar diseases. In animals, in addition to the scraping symptom, they may lose coordination, change temperament, lose weight, bite themselves, hop about like a rabbit, and be so startled by loud noises that they go into a falling down fit. Humans may lose coordination, as well as control of their eyes, hands, and voices. Slowly, they may lose the ability to stand and walk alone, or even to sit up. The inability to swallow and incontinence are other symptoms. The person may move in jerking motions and exhibit dementia and fits of uncontrollable laughter and depression. Bovine spongiform encephalitis ("mad cow" disease) and New Variant-CJD (Creutzfeld-Jacob Disease) are prion-caused diseases. According to scientists, since prion proteins are a normal part of the body, the immune system will never recognize them as harmful and so will not respond to the growth and development of the disease they may cause. Note that scrapie is a disease that had been around for

(426) hundreds of years, was on its way to being eradicated, and had never been known to infect humans ... until researchers made the connection that it was likely that scrapie had given rise to bovine spongiform encephalopathy ... and BSE to a new human disease of the brain—New Variant Creutzfeld-Jacob (nvCJD) disease. You may recall the panic in the British news when UK herds had to be destroyed and even tourists had to dip their shoes in disinfectant before leaving certain places in the countryside. The story of this disease is not over as we track its possible path to other countries and try to determine what this disease will do next, and where.

427

Just when we thought all we had to fear were "mad cows," a 1977 incident showed that CJD, Creutzfeld-Jacob disease, can be passed from person to person via surgical instruments. A woman being treated for epilepsy had a steel electrode inserted into her brain. Even though the instrument was thoroughly sterilized, after it was used on a 23-year-old woman and a 17-year-old boy, all three died, with CJD being the most obvious cause.

428

In February 2002, a new five-in-one
vaccine called Pediarix, developed by
GlaxcoSmithKline, combined diphtheria, pertussis,
tetanus, hepatitis B, and polio vaccines into one
shot. Some parents and doctors hailed this
improvement as easier on the child.
Some people worry that the combined shot may
overwhelm a child's immature immune system.
However, experts say that giving multiple vaccines
in any combination and from the same syringe is
perfectly safe. Today, babies generally get 15-20
shots in their first year of life. Some doctors see
the new combo shot as a way to get those
immunizations out of the way to make time for
childhood flu shots and possibly other vaccines.
While some adults suspect a connection between
vaccinations and diseases such as autism, many
doctors claim that people today don't realize that
without proper vaccinations, old diseases such as
measles and mumps could rage through a
susceptible population.

429

Scientists suspect that the
human papilloma virus is the culprit in some
forms of mouth cancer.

430

The Poetry of Germs

Germs may be a source of pain, fear, and even beauty. Some pathogens have inspired poetry. Ronald Ross, an Indian-born Englishman, won the 1902 Nobel Prize for determining that mosquitoes were the carriers of the germ that produces malaria. In a book recounting his discovery, he penned the following poem:

This day relenting God
Hath placed within my hand
A wondrous thing; and God
Be praised. At His command,

Seeking His secret deeds
With tears and toiling breath,
I find thy cunning seeds,
O million-murdering Death.

I know this little thing
A myriad men will save.
O Death, where is thy sting,
Thy victory, O grave.

431

There are approximately 380
species of anopheline mosquitoes around the
world, but only about 60 are able to transmit
the parasite that causes malaria. People
once believed that malaria came from bad air
(mal-air-ia), especially the foul air of swamplands.
There are several kinds of malaria, some milder
than others that can hang around in a human for
years if untreated. Other types, such as cerebral
malaria, which goes from bad to worse until it
causes coma and death, are far more severe.
In some countries, babies get bitten by so many
mosquitoes that they develop immunity to some,
or even many, forms of malaria, but not
necessarily all, so that a simple trip to a
neighboring village can end in
infection, illness, and even death.

432

Symptoms of Yellow Fever include fever, muscle
pain, headache, shivering, loss of appetite, nausea,
and sometimes, vomiting. If the disease
progresses, expect higher fever, jaundice, stomach
pains, bleeding from the mouth, nose, eyes and
stomach, including blood in the vomit and stool,
kidney failure, and possibly, death.

433

In January 2001, a healthy 18-year-old
boy in Mansfield, Texas died of pneumonia caused
by a S. aureus infection in just a few days, when a
nurse at the correctional institute where he was
housed failed to send him to a hospital.

434

During World War II, the Japanese operated Unit
731, a biological warfare research center in
Manchuria where they intentionally exposed at
least 3,000 Chinese prisoners of war to plague,
anthrax, and syphilis. Their methods included
tying prisoners to stakes and dropping biological
weapons on them from airplanes. Later, to
determine the results of these infections,
many prisoners were dissected without the
use of anesthetics. Plague-infected rats were also
grown at Unit 731 and then released in China,
where it is believed that they may have
killed at least 30,000 people.

435

Certain strains of the bacterium
Enterococcus faecium are resistant to almost all
of the more than 100 antibiotics doctors have
available in their germ-fighting arsenal.

436

During World War II, the United States formed the War Research Service which examined the possibility of using botulinum toxin and anthrax against the Germans in case they launched any biological attack against America.

437

In 1994, the same Aum Shinrikyo sect that introduced Sarin gas into the Tokyo subway system, also sprayed botulism from the back of a truck as it drove around town, and sprayed anthrax from the top of a tower the next year. Fortunately, only some birds were killed.

438

In April 2001, talk show host Rosie O'Donnell almost died after cutting her finger with a fishing knife and getting a multidrug resistant S. aureus infection. She described her hand as being so swollen that it looked like "a kid's bright red baseball mitt." Multiple surgeries were required to get rid of the dead and infected tissue and decontaminate the wound.

439

The American Type Culture Collection in Rockville Maryland is a non-profit company that contains America's largest collection of cultured diseases. It sends out at least 150,000 cultures around the world each year. In 1997, a man was caught pretending to be a researcher and ordering three vials of plague germs. (He was found guilty of mail fraud and sentenced to 200 hours of community service—presumably not in a hospital?) In 1995, the PrimeTime television show claimed that American Type Culture had sent salmonella to a cult and a strain of human-infecting anthrax to Iraq.

440

In the U.S., S. pneumo (Streptococcus pneumoniae) is believed to cause 500,000 cases of pneumonia, many in children, each year. Around the world, 1.2 million children die of S. pneumo infections annually. Today, 45% of S. pneumo strains are penicillin resistant and are growing resistant to the antibiotic erythromycin.

441

Some researchers believe that there is a connection between early rotavirus infection and the later development of diabetes. Who gets a rotavirus? Almost all children do in those years before they start school. Rotavirus infections are the major cause of diarrhea in young children. In developed nations, this means that 50% of the kids admitted to hospitals are suffering from rotavirus dehydration. In countries where such treatment is not readily available, 20% of infant deaths are from gastroenteritis caused by the rotavirus.

442

In tests of chickens intentionally infected with adenovirus-36, the birds put on up to 75% more fat than uninfected birds. Scientists wonder if viral infections could be a factor in some obesity. Tests show that overweight people are more likely than thin folks to have had an infection. But don't go off your diet, yet!

443

A "pinch" of dirt contains around one billion germs.

444

You can get infected with anthrax by getting it on a cut in your skin, eating contaminated meat, or by inhaling spores. Symptoms begin in one to five days. An anthrax skin infection (cutaneous) kills one in five people. Inhaled anthrax (inhalational) kills victims in about 48 hours. Anthrax not only releases a toxin that attacks the body, but also turns off the immune system. A vaccine works against the skin form of the disease; antibiotics help only if taken within hours of the infection. In November 2001, Americans experienced a bio-warfare event when anthrax-infected letters were sent through the mail to various media operations and governmental offices. It was not clear whether this was the doing of domestic or foreign terrorist(s). The illness and deaths that followed, although limited, alarmed many Americans, who did not anticipate such an attack.

445

Around one trillion germs form a protective layer on our skin. Tens of trillions call anywhere from our lips to our anus home. Their jobs include protecting us from more harmful microbes, helping digest our food and other tasks.

446

Germs are very clever and full of tricks.
After being exposed to antibiotics, they are often
able to develop resistance that lets them make
their cell walls impermeable to antibiotics.
Some create tiny pumps that actually vomit the
antibiotics out of the cell! As one scientist has
said, "The bugs are getting stronger—and
they're getting stronger faster."

447

According to some reports, of the 50 million
pounds (22,679.6 metric tons) of antibiotics used
in the U.S. each year, almost half were given to
animals. The small amounts given to animals
allow bacteria to become resistant to those
drugs. Such resistance is then passed on to those
who eat undercooked meat.

448

Drug resistance is being spread everywhere:
on the farm; in hospitals; in places where people
are crowded together, such as prisons,
military barracks, college dorms,
and even day care centers.

449

Around 1/3 of the people in the world are
infected with tuberculosis due to early childhood
exposure to the TB germ. Most of these
"carriers" live their lives without ever exhibiting
symptoms. However, once infected with some
other serious pathogen, such as AIDS, their
compromised immune system allows the TB
disease to then also develop in their bodies. In
other words, as AIDS has spread, so has
tuberculosis. And as TB is treated with
antibiotics, it grows ever more resistant to
antibiotics used against it.

450

The oldest known inhabitants of
the planet Earth are bacteria; they have
lived here at least 3.5 billion years, always killing,
always mutating, always adapting.

451

Think about it: without germs, the earth
would be filled with the wastes of all living
creatures since time began, as well as the
corpses of dead ones!

452

Antibiotics can kill you. In the gravely
ill, heavy antibiotic use can caused increased
diarrhea, which flushes vital nutrients out of the
patient, who can literally die as a result of the
antibiotic treatments.

453

"Scalded skin syndrome" is a condition in
newborns. Created by A. aureus toxins, the
syndrome causes the top layers of the skin to
exfoliate, exposing the sensitive underlayers
to infection.

454

Bacterial spores can survive for two hours in
boiling water. Even in an inhospitable
environment, they can live for 20 years. Four of
the deadliest diseases known to man are caused
by bacteria that produce spores. These include
anthrax, botulism, gangrene, and tetanus.

455

Chinese and African folk
doctors once used moldy soya beans
as the first antibiotics.

456

S. aureus causes more types of infections
in more people than any other bacteria.
It is the number one cause of hospital infections
in the world. At least 9 million Americans are
affected by S. aureus each year.

457

Cephalosporins, the most widely used class of
antibiotics in the world today (for pneumonia,
meningitis, middle-ear infections, and others),
were first discovered in 1945 in sewage water off
the country of Sardinia. Because human waste is
filled with bacteria, scientists suspected that's
where they would find natural antibiotics that the
gastrointestinal germs used to kill one another.
They were right!

458

Although Alexander Fleming, creator of
penicillin, warned that overuse could create
resistance to the drug by germs, by 1945 the
antibiotic was sold over the counter without a
prescription and put in cough drops, throat
sprays, soaps, mouthwashes, and even drinks! By
1948, 59% of S. aureus strains found in American
hospitals were penicillin resistant.

459

It was 1952 when some Japanese scientists first noted spontaneous multidrug resistance. Up until that time, microbiologists believed that even if mutations occurred, the use of two drugs in combination with one another would almost certainly zap any germ. Wrong!

460

The CDC (Centers for Disease Control and Prevention) in Atlanta, Georgia was created in 1946, primarily to fight malaria.

461

Drug companies, seeking natural antibiotics in soil, have often relied on missionaries in far-flung locales to provide samples of foreign dirt. In their search, they often identified potential antibiotic candidates, but had to learn what the organism fed on, for to develop the antibiotic on a large scale, the germs would have to eat well. The bug known as M4 3-05865, which tested for high antibioticity, was eventually shown to prefer a diet of milk, sugar, and Brer Rabbit brand molasses!

462

So grateful were people around the world following Louis Pasteur's development of a vaccine for rabies that many people, ranging from a mailman to the emperor of Brazil, donated funds to build an institution where such work against disease could continue—the Pasteur Institute in Paris, France.

463

Necrotizing fasciitis can destroy cells (eat your flesh) at about one inch per hour.

464

The EIS (Epidemic Intelligence Service) was established in 1951 during the Korean War era when there was fear of biological warfare. EIS turned into an emergency team which responded to epidemics of all kinds anywhere in the world. Their many accomplishments have included helping to eradicate smallpox, solving the Legionnaires' disease mystery, identifying the rare form of pneumonia that evolved into the AIDS epidemic, and many others.

465

For more than 50 years, warnings about antibiotic resistance—and the possible eventual end to the "antibiotic era" which has meant so much to mankind—has been anything but headline news, even in the medical community. Perhaps this dire potential seems more troublesome when you realize just how many ways germs can connive to survive—their whole goal in life. Yes, germs can mutate. But they have more in their bag of tricks than we might expect. Just a few tools of their trade include conjugation (what scientists call "bacterial sex"), transduction, and transformation. If that sounds like germs are quick-change artists, that's an apt analogy. DNA called plasmids can carry "R factors," genes that can convey antibiotic resistance. When one bacterial cell hooks up with another (conjugation), these R factor-carrying plasmids can jump from one cell to another. Like little SUVs let loose on a "spaghetti junction" interstate, transponsons (even smaller DNA) can get into a cell in several ways—even on the back of a virus!—via a process called transduction. And in the amazing process known as transformation, a cell could pass this DNA right out into the environment . . . and another cell can pick it up, as has been described, like a pass in a football game. No matter which method, all along the way

(465) bacteria are sharing ever-increasing antibiotic resistance to drug after drug. Note: Far more than once, scientists have speculated that the odds of any germ being able to do "this" or "that" and thereby become resistant to one, many, or all antibiotics was so infinitesimally astronomical as to be virtually impossible. However, never to date has that logic proven to hold true. Germs, it seems, are wily, creative, and determined beyond measure. Perhaps one day, it's the germs that will win a Nobel Prize for their amazing capabilities. No matter how much we might dismay over and even despise their frustrating talent at being able to eventually thwart all antibiotic comers—that's what germs keep doing.

466

Pristinamycin is an antibiotic created from the Argentinean fungus, Streptomyces pristinaspiralis. A natural drug of around 20 molecules (quite complex), one scientist nicknamed it "chicken soup." It is sold under the name Synercid.

467

The cost of getting a new drug through the required three-phase FDA (Food and Drug Administration) trials is around $500 million.

468

When the scientist who first identified tampons as the source of toxic shock syndrome shared his findings, he was ridiculed for two decades before others believed him.

469

In 1999, a three-year-old girl died after drinking well water contaminated with E. coli at a county fair in upstate New York. The E. coli had gotten into the well via cow manure tainted rain runoff to a nearby stream and into the well.

470

In the spring of 2000, a three-year-old girl died after eating meat tainted with E. coli at a Sizzler steakhouse in Milwaukee, Wisconsin.

471

Starting in October 1995, one scientist spent an entire year in the study of the feces of 30,000 turkeys in the hope of better understanding drug-resistance.

472

It would be difficult to sufficiently express within the scope of this "leisure-reading" book the fear and horror many scientists feel about the onward march of antibiotic resistant drugs. The acronyms VRE, MRSA, VISA, and others are tiny representations of big-time worries in hospitals and other places where major and full antibiotic resistance is found. Like an ongoing serial mystery, researchers attempt to track the circuitous path that pathogens take to emerge impervious to even the newest or most powerful natural or man-made antibiotic treatments. One example to serve the purpose of elucidating the no resistance/some resistance/multi-resistance/full resistance microbial square dance is this: A 1997 Michigan study of 100 random Minnesota residents found resistance to the antibiotic called Synercid. The astounding thing was that not one of the people tested had ever even been treated with Synercid. Why? Because in 1997, the drug had not even been released to the market! So how did resistance occur? Scientists believe that these people, like many others, had obtained resistance as a result of eating meat that had come from animals which had been fed certain growth-promoting antibiotics. The resistance that germs in the animals had developed had been passed on to germs in humans. If this sounds like drugmakers were making miracle drugs that were doomed from the start, that would appear to be the fact in many instances!

473

In about one in 1,000 cases, Campylobacter infections cause Guillain-Barré syndrome, which includes increasing paralysis and often death. Children are most often affected.

474

Salmonella kills about 500 people in the U.S. each year. It generally does this when the bloodstream becomes infected, creating a condition called bacterimia. In 2003, drug resistant salmonella was considered a "serious and growing threat" to humans.

475

Between October 12 and 14, 1996, 19 elementary age school children in Cass County, Nebraska came down with diarrhea, fever, headache, nausea, and vomiting—the first documented outbreak of five-drug-resistant Salmonella (DT104) in the United States.
It was never determined whether the source was expired milk, an ill kitten the children had handled, a "show and tell" turtle, or some other source.

476

A Wheeling, West Virginia student conducted a science project in which she tested the drinking water in her town for common antibiotics. She went on to win an international science prize for her findings of low to high concentrations of penicillin, tetracycline, and vancomycin in the local waters. The conclusion was that water filtration plants had not removed the antibiotics found in home and hospital sewage. Resistant germs were everywhere in these "drugged" waters!

477

In 1997 in Franklin County, Vermont, a family and their farmhands who drank fresh (unpasteurized) milk from their sick cows (or even handled the sick animals) began to fall ill with multidrug resistant Salmonella. Doctors were stunned by the animal-to-human transmission of the disease and had to work fast to save the lives of those who had become deathly ill.

478

Bugs and drugs: Polymyxin is a drug that kills germs... and people. Often used as a "last resort" drug, it can cause ulcers in the kidney and kidney failure, killing 25% of patients it is used on.

479

MRSA (methicillin-resistant S. aureus)
is nightmarish enough, but in 1998 and 1999,
doctors were dismayed by a devastating string of
pediatric MRSA cases, especially since some of
the children had spent no time in a hospital, the
typical source of this insidious infection. The
fatalities included: a 7-year-old African American
girl in urban Minnesota, who died after five
weeks; a 16-month old Native American child
from rural North Dakota, who died within two
hours of coming to the hospital; and an 11-year-
old girl, who also succumbed to multi-organ
failure, then death, as a result of an infection that
today's wonder drugs could not faze.

480

The highest rates of penicillin resistance are in
wealthy suburbs. Why? Possibly because the
parents there can afford to take their children to
the doctor anytime they get sick, where doctors
often treat them (egged on by the insistent
parents who do not want to see their children
suffer) with unnecessary antibiotics.

481

When all is lost, what can you do?
In a word—immunogammaglobulin. This pool of antibodies from the immune systems of 1,000 human donors has been available since the late 1980s. It can be helpful when a person's immune system needs a lifesaving boost. Because it is drawn from such a wide variety of volunteers, it is likely that antibodies to most germs exist within the immunogammaglobulin. Because it is expensive, and other measures would certainly have already been tried and failed, this course of treatment is given only as a last resort, especially for diseases where antibiotic resistance is discovered to be a factor.

482

"Hospital gangrene" was a frightful strep infection during the 1800s. During the Crimean War, the French threw 60 infected men overboard in hopes of keeping the entire crew of a hospital ship from being infected.

483

An Egyptian bas-relief from 1500 B.C. depicts a priest with a shriveled leg, probably from a polio infection.

484

It was once called a "dumb" bug because it could not seem to catch on to how to become resistant to antibiotics. Streptococcus pneumoniae is the germ that causes bronchitis, sinusitis, pneumonia, and in children, more than 6 million earaches each year. S. pneumo is also the biggest cause of acute bacterial meningitis (which causes a potentially fatal swelling of the membrane around the brain) in children. At least one million child deaths around the world annually are attributed to this so-called "dumb germ." Moreover, in recent years S. pneumo, due to the massive amounts of antibiotics prescribed for respiratory problems, has gotten smart virtually overnight and has now grown resistant to penicillin and many other drugs. Between 10-25% of hospital patients die from pneumonia. Children who contract ear infections so constant and severe that tubes may have to be inserted in the eardrum to drain infectious fluid, sometimes also face medical nightmares requiring surgery and even resulting in a loss of hearing.

485

A 13th century Arabic manuscript depicts a mad dog, presumably sick from the rabies virus, biting a man.

486

"Finger of Faith"—that's what one woman calls the tip-clipped, deformed finger she was left with after a bout with an infectious disease known as necrotizing fasciitis, nicknamed the "flesh-eating" disease. It started with a simple puncture, probably from a spider bite, in her left index finger. In a few hours, she felt like she was coming down with the flu and had intense pain in her finger. Next, her finger began to change colors. In spite of a host of excuses that could have kept her home, she went to the emergency room that night where doctors operated on her then-blackened finger. Doctors told her that if she had not come in, she would have died in the night. Even at the hospital, she stayed an hour from death's door as doctors treated her 22 more days before she went home. (A previous patient with the same disease had gone home with no arms and no legs.)

487

Many peoples have worshipped various gods or goddesses of smallpox. Just a few include T'ou-Shen Niang-Niang and Pan-chen (China) and Shitala Ma (India).

488

Swimming, anyone?: A single spoonful of seawater can contain more than 1 billion viruses.

489

Necrotizing fasciitis is no joke, as you might imagine from something also called "flesh-eating" disease. A certain strep germ that usually lives in the throat can cause strep throat, scarlet fever, or kidney disease. When this strep gets on the skin, it can cause something as simple as impetigo or something as potentially deadly as necrotizing fasciitis. Sometimes, this germ can find its way into the body where it is activated into infection by something as ordinary as a bump on your funny bone. That's pretty much what happened to one man who bumped his elbow on a table during a morning meeting. He went on about his business until he felt feverish and went home where his wife discovered his arm was bright red with red streaks radiating outward from the point of the initial bump. They headed for the emergency room where doctors operated to remove dead and dying flesh and filled the man with massive amounts of antibiotics to stop the infection. He was told that had he waited until morning to come in, he would have certainly died.

490

Jim Henson, creator of Kermit the Frog, Miss Piggy, the Cookie Monster, and other Muppet puppets of Sesame Street and The Muppet Show, died at age 53 from what was essentially necrotizing fasciitis ("flesh-eating" disease) of the lungs. It began during a round of golf where he complained of flulike symptoms. By the next day he was coughing up blood with a tumor the size of a softball in his lungs which killed him via toxic shock syndrome.

491

In the "you can't win for losing" category: The human immune system is designed to attack toxins at a particular site of infection, not attack on many fronts. When something like necrotizing fasciitis releases toxins that get into the bloodstream and affect much of the body, your poor immune system goes into overdrive trying to fight multiple battles throughout your body. This attempt to help save itself creates a "cytokine storm" of hormones which flood the body in an overreaction that causes widespread tissue damage. Eventually (even though antibiotics may have stopped the original infection), blood pressure plummets and vital organs such as the kidneys, liver and heart shut down to create death by toxic shock.

492

Oh, my aching algorithm! Gideon, a software program, diagnoses infections. How? You enter certain variables, such as symptoms, etc., and the program digests them to come up with 5 potential germ candidates that could be the cause, from least to most likely.

493

In 1994, the future premier of Quebec, Canada, Lucien Bouchard, went to the hospital with a pain in his leg that doctors believed might be a blood clot. Instead, he was diagnosed with necrotizing myositis, a variant of necrotizing fasciitis. Much of his leg had been "eaten up," and the infection was rapidly spreading to his abdomen. Quickly, the doctors amputated part of his leg and pumped him with immunogammaglobulin (IVIG) in hopes of stoping the infection without starting a cytokine storm that could overwhelm his immune system. Fortunately, with such quick diagnosis and action, they were successful.

494

Around 20 people, many of them children, died in a 1998 epidemic of a deadly form of strep infection in Austin and Houston, Texas.

495

In the late 1990s, a mini-epidemic of necrotizing fasciitis affected nine people in Missoula, Montana. Nearly half of those infected died and most of the others had parts of their arms or legs amputated. Another mini-epidemic of the "flesh-eating" disease occurred in Chicago in 1999, infecting 15 people. Random instances of the disease also killed: an eight-year-old boy in Queens, New York, who apparently had gotten the disease as a result of chickenpox; two prostitutes in San Francisco died of the disease after sharing needles while doing drugs; and, an elderly Florida man with a small cut on his arm who died quickly in spite of treatment. To this day, doctors have little to go on about when and where and why the disease strikes. In spite of some correlations between colder climates, HIV, cancer, diabetes, alcohol abuse, and chickenpox, the disease seemed to care little whether you are old, young, healthy, male, female, rich, poor, a Yankee or Southerner. I guess we all taste equally good to these horrible, hungry germs.

496

If a smallpox patient lives long enough, the whites of his eyes turn solid black.

497

Germs known as Gram-negatives (Pseudomonas aeruginosa, Klebsiella pneumoniae, and Acinetobacter baumannii are examples) cause infections of the urinary tract, gall bladder, bile ducts, kidneys, bloodstream, lungs, and many other miseries. In fact, they cause 40% of the 2 million infectious diseases in America each year.

Gram-negative germs are everywhere, literally! Not just in our gut and on our skin, they are also found in dirt, water, and on all things in a hospital like sheets, bedrails, stethoscopes, and the hands of doctors and nurses. They love vegetables (and cause them to rot, which should give you some idea of their abilities) and plants (so much so that flowers are prohibited from the rooms of patients on chemotherapy.) Not only are Gram-negative germs unfazed by penicillin, that antibiotic actually fosters the spread of those germs by killing Gram-positive infections, leaving extra room for Gram-negatives to multiply. Gram-negative germs are especially the enemies of anyone on dialysis machines or ventilators since the germs use these as transportation into the bodies of the already seriously ill.

498

Bedeviled by bugs: In 1995, a 16-year-old boy in Madagascar was diagnosed with malaria and treated. However, instead of improving, he grew much worse with high fever, delirium, and a pus-filled growth in his groin. The ugly bulge was a bubo, evidence that he did not have malaria, but, instead, bubonic plague. The bubo was punctured and drained. Then the patient was treated with antibiotics usually used to treat the disease. None worked; all were resistant. Needless to say, researchers were terrified. Although bubonic plague was considered an awful disease of the distant past, the truth is that there are still occurrences (around 3,000 annually), even epidemics, around the world each year. For example, more than 6,000 people in India were infected with Yersenia pestis in 1994, following an earthquake which may have stirred up buried bacteria. Because of antibiotics, only 50 people died. But what if new strains prove resistant to antibiotics? So fearful were scientists of this strain, that they did not even want to send it to labs for additional study. In fact, the CDC intentionally requested samples of the strain just to make sure they were turned down!

499

Bioplague: Bubonic plague is one of the top four infectious diseases that scientists worry may be used for bioterrorism. Although turning plague germs into weapons is a complex task, it is not impossible. According to a high-ranking Soviet official (once deputy chief of their germ warfare agency, Biopreparat), drug-resistant plague has already been produced in no less than 40 labs in 15 Russian cities. Both naturally resistant strains and genetically engineered strains had been developed. Naturally occurring strains are actually easier to "weaponize." At one time, according to this official, Biopreparat had 20 tons (18 metric tons) of plague in storage.

500

Hope from nature: In this worrisome era of antibiotic resistance, researchers continue to search for natural antibiotics. Their research has uncovered some hopeful prospects, especially human peptides which can kill bacteria. Horseshoe crabs, some insects, cows, frogs, and even dogfish shark have peptides— more than 500 types.

501

The nose knows: In early 2003, tests
on nine victims of Cruetzfeldt-Jacob disease
found that all had defective protein particles
linked to the illness in their nasal passages. Tests
of a similar number of people with other types of
nerve disease showed none of these particles in
their noses. Until now, the only definitive way to
prove that someone has actually died of CJD
is a brain biopsy after death, which is seldom
done because it is so dangerous. The disease,
which has no cure, leaves the brain so full of
holes it looks like a sponge.

502

I'd be scared too! During the early 2003 Ebola
outbreak in the Congo Republic, more than 59
people died and at least 70 were infected as the
outbreak spread. Efforts to contain the outbreak
were stymied by villagers' fear of not only the
horrendous symptoms of the disease, but also of
the health workers in the oversized spacesuits,
from which the terrified villagers fled.

503

The hills are alive ... with germs: Although
bacteriophages (also called phages) are the latest
darlings in the germ-killing world, their use dates
back much earlier. During World War II, the
Pasteur Institute was a leader in phage research.
With the fall of France, the Nazis occupied the
Institute hoping to make phages to treat infections
on the battlefield. Adolph (the "Big Germ") Hitler
is said to have used the code-name "Edelweiss" for
his plan to seize Russia's Eliava Institute's phages for
the same purpose, but was defeated. However, the
Russian army did use phages successfully in the
war, against their soldiers' dysentery and gangrene.
Unfortunately, their success with phages was little
known, and therefore, not studied by the West for
many years. But in the Soviet Union, phages were
manufactured in tablet and liquid form and sold to
prevent gangrene, dysentery, and many other
medical problems. At one time, one plant
produced two tons (1.8 metric tons) of
phages each day.

504

Epidemiologists estimate that smallpox killed
around one billion people during the last 100 years.

505

Two scientists doing research on phages obtained their samples in the university sewage system and harbor in Baltimore, Maryland, as well as in the Chesapeake River. Today, phages are genetically engineered. One is even called Rambo because it is so powerful.

Man, Alive . . . There's More Than 505! So, let's continue in Germ Land, shall we?

506

In February 2003, the CDC was scheduled to receive $268 million for much-needed repairs and new facilities. After all, the place that is supposed to be the crown jewel in protecting us from germs has long suffered from greatly-deteriorated buildings and laboratories dating back to the 1950s.

507

On you huskies! Alaska's almost 1,000-mile-long (1,609 meters) annual Iditarod Trail trek competition was begun in 1925 by mushers who risked their lives to carry medicine through a blizzard from Anchorage to Nome in record time to prevent a diphtheria epidemic.

508

Only a plane ride away: In 1998, a man from the Ukraine flew to New York City. He coughed all the way. Unfortunately, no person or system stopped this actively-infected with tuberculosis man from entering the country. He was so ill that he soon checked himself into a health clinic in Pennsylvania. When he was treated for the disease, it was discovered that the strain of TB he had was resistant to six types of antibiotics. Health investigators tracked down around 40 passengers who sat near the man— and all tested positive for TB.

509

Researchers believe that it only takes as few as three to five virions, or particles, of smallpox virus to be inhaled to become infected.

510

And for your next vacation...
The National Parasite Collection is located in
Beltsville, Maryland. It includes tapeworms large
enough to live and grow inside whales. Or visit
the National Institute of Pathology in Forest Glen,
Maryland, a repository for a century's worth
of tissue samples, sorted and retrieved
by robots.

511

In 1989, a man attended the funerals of
both his parents in Nigeria, then boarded an
airplane and flew to the United States. He ended
up in a clinic in Chicago complaining of fever and
a sore throat. The doctors sent him home with
antibiotics, but the man soon died—of Lassa
Fever, a deadly hemorrhagic disease.

512

From 1964-1996, Texas recorded 964 cases
(the largest number of any state) of St. Louis
encephalitis, a disease acquired through the
bite of an infected mosquito.

513

If you can't fight 'em, join 'em. Since viruses attack specific cells, doctors have been able to send new genetic information to sick cells by using a virus as a "messenger." Such methods have been used to replace the defective gene that causes cystic fibrosis in children with a good gene. Researchers have high hopes for the use of genetics to combat germs.

514

Breast-feeding is recommended because mothers' milk is rich in antibodies, giving nursing children protection against diseases while they build up their immature immune system.
In most countries, powdered formula is used with no ill effects. However, in countries with contaminated water supplies, formula can become infected with rotavirus, which kills about one million children each year.

515

In early 2003, a news report in New Mexico stated that with the use of Doppler radar, scientists would be able to predict the next Hanta virus epidemic—down to a specific house!

516

Many soldiers died of kidney failure
brought on by Hanta virus during the Korean
War. When the virus came to the United States,
it mutated, and instead, attacked the lungs.

517

In May 1996 on the Ivory Coast of Africa,
more than 100,000 people became infected
with bacterial meningitis in three months;
more than 10,000 died.

518

Uh, oh, ozone? Scientists from the University of
Texas are studying the relationship between a
thinning ozone layer and a possible suppression of
the human immune system. Likewise, scientists
are studying the relationship between the
greenhouse effect and the rise in malaria due to
the increase in mosquito breeding grounds.

519

Junin virus is found only in Argentina
and is carried by rats. It is similar
to Lassa, one of the hemorrhagic fevers,
and its death rate is high.

520

In 1990, a 22-year-old man was bitten
on the finger by a bat. Six weeks later,
he grew quite ill. His hand was weak and his
body would go rigid as he held his breath.
He could not swallow and began to hallucinate.
His neck, face, and mouth jerked in spasms.
He had a high fever, was disoriented, and drooled
constantly. In a few days, he fell into a coma
and died. Cause of death: rabies.

521

Pregnant women are warned not to
clean litter boxes because cats are the primary
carriers of the parasite that causes toxoplasma
infections, which in humans, can cause
miscarriages or fetal heart or brain abnormalities.
Now, sea otters off the coast of California have
been found to be infected with this same disease,
presumably from the runoff of cat waste reaching
the ocean water where they swim.

522

In 1996, three Ebola-infected monkeys
were quarantined in Alice, Texas by a company
that buys and sells monkeys for laboratory
research. During a required 31-day quarantine,
one monkey died and 50 were destroyed
as a safety precaution.

523

Salinosporamide A is a microbe found deep in the
muck of the ocean floor beneath tropical seas.
Tests show that it can inhibit cancer growth,
including cancer of the colon, breast, and lung.

524

Researchers believe that giant sea
slugs off the coast of California contain a protein
that may help create a vaccine for
non-Hodgkins lymphoma.

525

Scientists have discovered that a
substance in the tissue of the common sea squirt
(an ugly little sack-shaped animal) is effective in
treating breast and other soft-tissue cancers.

526

While Legionnaire's disease is
often connected with air conditioning units,
it can also thrive in hot water heating units,
and even in shower heads.

527

Some scientists believe that the Coxsackie B virus is the culprit behind juvenile diabetes and obsessive-compulsive disorder.

528

Scientists are exploring the possibility of treating rabies with interferon, a naturally occurring protective protein that the immune system produces to fight viruses. To date, no one has ever survived rabies. Doctors say that treating someone at the ending stage of this disease is particularly disturbing because patients suffer severe thirst, convulsions, and are hypersensitive to the slightest touch, often even unable to stand a sheet over their body.

529

Prior to the anthrax deaths in 2001, following September 11, there had been only 18 cases of inhalation anthrax in the past 100 years in the U.S., with the last reported case occurring 23 years earlier.

530

Although books and movies portray hemorrhagic diseases as virtually turning body organs into pools of blood, this is not quite true. While the circulatory system can be destroyed, creating the "red" whites of the eyes syndrome and bleeding from any and even every body orifice, vital organs do not tend to liquefy or turn to a bloody pulp. In fact, it seems a mystery to doctors how and why the organs stay so intact in the face of these dismal diseases.

531

One healthy young woman infected with Lassa Fever had the following typical symptoms near the end-stage of the disease: severe swelling of the head and shoulders so that she was virtually unrecognizable; decerebrate rigidity, a condition that freezes the patient in such a contorted position that their head and arms and legs extend backwards behind their body; and, lack of response to even the most painful stimuli. (This person survived and claimed no recollection of the course of her illness!)

532

Monkey B is a herpes virus that only causes a monkey to have cold sores. However, in humans it can cause a rabies-like disease which is almost always fatal. In the 1980s, a monkey handler died near Pensacola, Florida of the disease after handling an infected monkey. So rare is Monkey B that even most researchers forgot that it can kill humans. This poor man must have suspected something, for after his death a book turned to a page about Monkey B was found open on his desk. Alas, he had shared his fears with no one.

533

What were they thinking?
In an autopsy of one of the 2001 anthrax victims, no one had ever done a post mortem on anyone who had died of anthrax. Once the man's chest was opened, anthrax-infected blood began to pour from this body cavity and spill all over the floor. The examiners knew that once these cells encountered air, they would quickly start to turn into anthrax spores.

534

Some gloves worn during autopsies
or other work with infectious diseases
are lined with Kevlar, the same material
used in bulletproof vests.

535

Who says they don't have a sense of humor? One
infectious disease expert owns three cars with
tags LASSA 1, LASSA 2, and LASSA 3.

536

During an outbreak of Crimean Congo
hemorrhagic fever in Saudi Arabia, one scientist
made the connection between the disease and
possibly infected animals. Because local religious
customs did not permit the actual counting of the
number of animals you owned, one creative
researcher counted the animal feces left behind
each night in a compound to more or less
determine the statistics he was after!

537

The germs that cause the common cold are the
smallest virus particles found in nature.

538

Partly due to crowding and poor sanitation in some places, India has more new polio cases than any other nation. As the total eradication of polio from the face of the earth is chased through remaining hotspots, progress is being made. Despite a civil war, millions of Somali children have been vaccinated in the past few years; only seven new polio cases were reported in 2001. Around 7 million children in the Democratic Republic of the Congo were inoculated in July 2001 alone. In 1988, nine years after the last U.S. outbreak of polio, the World Health Organization (WHO) set a goal to eradicate polio worldwide. That same year, 350,000 people (mostly children) in 125 nations contracted the disease. By 1999, the incidence of new cases had dropped 99% to fewer than 1,000 cases in 10 nations. However, to completely eliminate polio, it is estimated that 10 million health workers will have to vaccinate 1 billion children. One obstacle is that oral vaccine—which must be administered to a child on three separate occasions to be effective—is used instead of a shot because the oral vaccine costs around 10 cents versus $1.00 for the inoculation.

(538) To know who has been vaccinated, health care workers write P on the door of a house; the child's thumb is also marked. Vaccinators must be brave and tireless; in 2001, three polio workers were killed in Somalia by gunfire or in car wrecks, and two were taken hostage. One downside of the oral vaccine is that it can cause outbreaks of the disease, although this is rare. Some people argue that the time and money spent on this goal would be better spent against malaria, AIDS or tuberculosis, which are more common in the developing world. One doctor, called "an incredible hero," operates on children maimed by polio for free.

539

The Secure Room at the CDC holds safes which contain the formulas for biological weapons capable of mass destruction. The patents on bioweapons created by American scientists are classified.

540

The exact cause of death in smallpox is not known.

541

In 1969, a man in West Germany (who had recently traveled in India, Pakistan, Turkey and other countries) came down with smallpox. For a while doctors had no idea what was wrong with him. A mass of pus-filled blisters covered much of his body including his genitals, the inside of his mouth, ear, and sinuses, his tongue, and, likely, his rectum, as well as on his face, arms, legs, and trunk. As the pustules grew and spread, the skin pulled away from the underlying flesh, effectively stripping the skin from the body until it hung baglike, and he was in great pain. As the pus threw off gasses, a miserable odor emanated from the suffering man. He lived to tell the tale. He also had been vaccinated—twice—but a scar had not formed on his arm, proving the vaccination had not "taken." Obviously.

542

In March 2003, a man was diagnosed with a smallpox infection in his eye after coming into contact with a person who had recently been inoculated during the military's vaccination program.

543

People who are immunized with the
traditional multi-prick shot against smallpox are
actually infected with the vaccinia virus. This
causes a single pustule that leaves a scar. If you
get the shot, it is believed to offer protection up
to four days after you have inhaled the virus.
(The incubation period of smallpox is 11-14 days.
If you get a shot after four days, it is too late to
provide any immunity.) If you were inoculated
against smallpox in the past, your immunity
probably began to diminish after about five years.
If your scar has disappeared, almost certainly
so has any immunity whatsoever to smallpox.

544

The many flavors of smallpox:
Mild=varioloid rash. Discrete ordinary
smallpox=creates separate blisters. Confluent
ordinary smallpox=blisters merge into sheets;
typically fatal. Hemorrhagic smallpox=bleeding
occurs under the skin; 100% fatal. Flat
hemorrhagic smallpox=the skin is not blistered,
but smooth and darkens until it looks charred
and slips off the body in sheets; also called black
pox; most common in teenagers; represents
3-25% of smallpox cases; fatal.

545

A pox on your house: There are two main kinds of poxviruses—those of insects and those of vertebrates. Just a sampling of the types of poxes: mousepox, monkeypox, skunkpox, pigpox, goatpox, camelpox, cowpox, buffalopox, gerbilpox, deerpox, sealpox, turkeypox, canarypox, pigeonpox, peacockpox, parrotpox, toadpox, mongolian horsepox, dolphinpox, penguinpox, kangaroopox, racoonpox, snakepox, crocpox, beetlepox, butterflypox, mosquitopox, caterpillarpox, and many more, including a pox called orf!

546

In 2002, the U.S. suffered the biggest reported outbreak of West Nile encephalitis in the world for that year. This included at least 2,500 reported cases in all but five states. It killed 230 people compared to nine deaths in 2001.

547

In December 2002, the first known case of a baby contracting West Nile virus from its mother before birth was reported.

548

It is estimated that one case of smallpox can lead to 1,000 or more outbreaks of the disease.

549

In November and December 2002, flesh-eating bacteria killed at least one recruit at the Marine Corp Recruit Depot in San Diego, California, made 100 others ill, and required the treatment of around 6,000 other soldiers who had possibly been exposed.

550

The world's last case of smallpox occurred on September 15, 1975 in Chittagong, on the eastern side of the Bay of Bengal. Ali Maow Maalin, of Somalia, is also said to have been the last case of natural smallpox, in 1977. Of course, so is Janet Parker of Birmingham England, in 1978.

551

Several cases of smallpox were once diagnosed by an 8-year-old girl who reported them to a health care worker and collected a reward of $62 from WHO.

552

Workers who helped eradicate
smallpox included a man who rented an elephant
to ride through villages encouraging people
to get vaccinated, the Lions Club, Rotary
International, Jerry Garcia's doctor, Wavy Gravy,
Baba Ram Dass, and Steve Jobs.

553

In 1975, at least 75 labs had frozen stocks of
smallpox virus; it may be unknown today how
many laboratories have variola in their freezers.
Officially, variola is located only in the Maximum
Containment Lab at the CDC and in Russia. The
WHO does not allow any lab to have more than
10% of the DNA of smallpox and no lab is
allowed to experiment with smallpox DNA.

554

In 1991, WHO destroyed most of its
200 million doses of frozen smallpox vaccine in
Geneva, Switzerland to save the $25,000 a year it
cost to keep it frozen. Today they have only
500,000 doses of smallpox vaccine.

555

It is said that the smallpox at the CDC takes up about the same amount of space as a basketball. It is kept in a freezer. No one knows what this freezer looks like or where it is stored. It is on wheels and can be moved around. It is also covered with chains and many padlocks and is bolted to the floor or walls. The freezer, and possibly several fake counterparts, and perhaps even a freezer A and freezer B (so that the smallpox is divided or "mirrored" so that if something happens to half of it, the rest is ok), are all heavily alarmed. These alarms, if sounded, hail armed federal marshals.

556

Only 1% of the United States' budget for health care is spent on disease prevention.

557

Tissue issue: In Charlottesville, University of Virginia students once volunteered to "catch a cold" in exchange for $350 so that they could be studied by a virologist seeking a cure.

558

Famous American author, Upton Sinclair,
once wrote a letter to U.S. president
Theodore Roosevelt urging him to look into the
unsanitary conditions in Chicago meat-packing
houses. The writer claimed that the beef was
tainted with tuberculosis, and that poisoned rats
were ground up and put into the meat.
His cry for help (and his exposé book, *The Jungle*)
disclosed the dangerous conditions which could
have led to illness for many Americans.

559

Wilma Rudolph was diagnosed with polio as a
baby. She overcame the disease and went on to
become the first U.S. runner to win three
Olympic gold medals, in the 1990 Games.

560

On the Great Plains, pioneers suffered greatly
from diseases. Cabins could turn into instant
"hot zones" once malaria or cholera struck. In
the winter, pneumonia spread quickly in crowded
soddies or cabins. The sick, if treated at all, were
treated with onion poultices, "Indian Primp" tea,
and boiled corn tucked next to their bodies.

561

One of the dreaded plagues on the Oregon Trail was cholera. Sometimes the death toll from this disease was so high that children had to assume the tasks of driving the wagons, hunting food, making decisions, and even fighting Indians! Children of the era especially suffered from "childhood diseases" such as measles and scarlet fever. Typhus and smallpox were also rampant, spreading quickly in tight quarters such as winter encampments or aboard steamboats. "Mountain fevers" caused by tick bites also prevailed. With little or no medical aid available, the misery was intense and the outcome often death.

562

In the Massachusetts Colony, Cotton Mather promoted inoculations against the smallpox epidemic sweeping the city. Only one of Boston's ten doctors agreed to try this solution. Using a sharp toothpick and a quill, Dr. Zabdiel Boylston vaccinated his six-year-old son and two slaves with pus from a smallpox patient. All three developed mild infections, which left them immune from the deadly disease.

563

Infectious diseases are the third leading
cause of death in the world.

564

Most organisms that infectious disease specialists
work on today were unknown 25 years ago.

565

Anthrax can kill so fast that
a dead hippopotamus standing upright was found
in Zimbabwe. Anthrax had killed the
animal as it was walking!

Ok, let's keep those germs comin'. . .

566

One emergency medical planner is so convinced
that a bioterror event will happen at some point
in New York City, that he keeps a strip of
reflective tape on the roof of his car so that in
the event of a bio-disaster, emergency police can
spot him from a helicopter.

567

Tuberculosis kills more women and girls around the world than any other infectious disease. More than 900 million females, one-third of all the women and girls in the world, are infected with TB. More than one million of them die of the disease each year. Even though there are drugs to stop this disease, in many countries, women and girls get little or no medical care.

568

People who attend church have stronger immune responses to infections.

569

In March 2003, the response to voluntary inoculation against smallpox had been so low that one major newspaper reported only seven health care workers were ready to respond to a smallpox epidemic, if one occurred, in New York City. The goal is to vaccinate hundreds of thousands of health care workers and volunteers to serve on Smallpox Response Teams in the event America is attacked.

570

Each sample—more than 30,000 by
the time it was over—of anthrax from the 2001
attacks was technically criminal evidence and had
to be put in a federal evidence tracking file folder
with more than 100 sheets of paper documenting
the chain-of-custody of the sample.

571

During the 2001 anthrax attacks,
one infected man went to an emergency room in
Maryland, but was sent home because doctors
did not recognize what was wrong
with him or that he was dying.

572

Henrietta Lacks is probably the most famous
woman you never heard of. She died in
Baltimore, Maryland in 1851 of cervical cancer.
Her cells—known as HeLa cells—are widely used
in medical research in cell-culture growth and
testing of microbes. God bless Henrietta!

573

Scientists have created an algae that makes
antibodies that target the herpes virus.

574

Not exactly a beautiful day in the
neighborhood: On September 11, 2001,
as scientists were inside working on smallpox-
infected monkeys, the entire CDC was ordered
to be immediately evacuated in anticipation of a
possible attack. The scientists took time to lock
and chain the open smallpox freezer, then had to
scramble through emergency procedures
and crash their way out of the
Maximum Containment Lab.

575

Ken Alibek has a doctor of sciences
degree in anthrax. He was once the head of the
Soviet Union's bio-warfare production facility, at
that time, the largest in the world.

576

Scientists say that the ingredients
needed to make anthrax (except for the anthrax
itself, presumably) can be found at
local home improvement stores.

577

Superpox: Scientists worry that bioweapons can create infectious agents that can "crash through" the vaccine that is supposed to protect people from that agent.

578

In 1999, a jar that contained a child's arm was found in the basement of the Indiana University School of Dentistry. The jar was labeled M243 Smallpox.

579

If you want to create your own designer virus, you need the four-volume ring binders nicknamed the "cookbook"—Current Protocols in Molecular Biology. Then, you can order strains of micro-organisms from someplace like American Type Culture Collection in Manassas, Virginia.

580

In early 2003 the Food and Drug Administration announced plans to require a new warning on antibiotics, pointing out that overuse makes them less effective.

581

On March 20, 2003 war against the Iraqi regime of Saddam Hussein was undertaken by the U.S. military and its allies. The goal—enforce the disarmament of so-called weapons of mass destruction that the Iraqi leader had been known to use against his own people, including sulfur mustard (SM). As the war raged on, some WMDs were indeed discovered and destroyed.

582

In April 2003, Canadian researchers reported that a powder made of freeze-dried egg yolks could be sprinkled or sprayed on meat, fruit, and vegetables to guard against common food-borne germs including E. Coli and salmonella. This tasteless powder, which remains active for a couple of hours, was said to be the antibiotic of the future.

583

It is believed that playwright William Shakespeare died at the age of 52 from typhoid fever.

584

In mid-March 2003, a mysterious, contagious, and deadly form of a pneumonia-like illness caused the World Health Organization to put out a rare emergency travel advisory. More than 150 reports of Severe Acute Respiratory Syndrome, or SARS, as the illness was initially called, occurred in just one week. Most of the cases were in Southeast Asia, with three deaths, including an American businessman and two people who had traveled to Canada from Hong Kong. A doctor believed to be infected was taken off a New York-to-Singapore flight in Germany and quarantined. A man traveling from Atlanta to Canada was also reported to have symptoms of the strange disease. Although the CDC immediately went on emergency status, they could not determine if SARS was caused by bacteria or a virus. Neither antibiotics nor antivirals appeared to have any effect on the disease. The symptoms of the disease include high fever, coughing, shortness of breath, headache, muscular stiffness, loss of appetite, confusion, rash, and diarrhea. It apparently is spread from person to person through close contact. Numerous medical personnel became infected after treating SARS patients. While no immediate restrictions

(584) were put on travel, travelers who did not have to visit southeast Asian countries were encouraged to postpone trips. The outbreak was not believed to be bioterrorism. It was also immediately unknown if the pathogen was something new on the world germ scene or if it was a germ that had mutated into a new more contagious, untreatable form. The big concern was that SARS seemed to be heading around the world via airplanes. After further examination, scientists suspected SARS of being a "paramyxo-virus" related to measles and mumps, of East Asian-origin, and associated with fatal lung infections. Later, it was determined that SARS was caused by a new corona cold virus. Toronto, Canada suffered an outbreak that worsened after a worker with symptoms was cautioned to go home but returned to work. Scientists predict that even after they determined the actual cause of the disease, any vaccine would be years away.

585

Extremophiles are microbes that thrive in particularly inhospitable habitats. Thermophiles live in boiling water. Psychrophiles live in frozen lakes beneath Antarctic ice fields.

586

Researchers are developing hand-held devices which could detect anthrax, mustard gas, and other bioweapons more quickly than current laboratory tests. However, it will be five years before this "works in one hour" sensor will be ready for use.

587

An increasing number of people are intentionally choosing to eat or drink germs. Prebiotic products are dietary supplements and dairy products that may prevent disease. You can drink a three-ounce serving of fermented milk stuffed with 10 billion Lactobacillus casei, for example. While antibiotics kill good and bad germs, prebiotics are normal good germs that live in your gut. In fact, they help keep harmful bacteria at bay. Don't think you've eaten any germs lately? Have you had a cup of yogurt?

588

Mycoplasmas are slow-growing, primitive bacteria that create symptoms such as fatigue, headaches, soreness, and joint pain.

589

Nanobes are microbes that are smaller than the smallest known bacteria. They do not need a host in which to live. Nanobacteria are 1/1000th of the size of regular bacteria. Because they are so small, nanobes can move around in cells and kill them. Nanobacteria were only discovered in the early 1990s. One of its secrets is that it can create a sort of "slime" to disguise its harmfulness to our bodies. In the meantime, it attacks the body, leading to plaque in our arteries, heart disease, kidney stones, eczema, liver cysts, breast calcification, psoriasis, dental plaque, rheumatoid arthritis, fibromyalgia, multiple sclerosis, Alzheimer's, and other diseases. There is no known natural enemy of nanobacteria.

590

Scientists believe that these are some of the major reasons some diseases are making a comeback: changes in food processing and handling; increased use of child care facilities; substance abuse; unprotected sex; changes in land use; urbanization of former tropical habitats; breakdowns in the public health system.

591

Too Pooped to Pop?
One doctor recommends this quick "self-help" check to see if the 400 or so known species of bacteria which inhabit your intestinal tract are doing their job: If your stools are large in diameter, brown, and float—that's good. If your urine is yellow at least once or twice a day— that's good. Be brave! Check it out!

Be sure to wash your hands early and often!

592

Scientists are beginning to suspect that the "vestigial" organs—tonsils, adenoids, appendix, and spleen—actually serve important roles in the human immune system.

593

Generic dishwashing soap is up to 100 times more effective than antibacterial soap in killing germs that cause respiratory diseases.

594

The only way to eliminate the accidental spread of "mad cow" disease through contaminated surgical devices is to destroy them after a single use. Since some of these devices cost thousands of dollars, hospitals are reluctant to follow this prescription for prevention!

595

Eat Good Food ... Some spices have been shown to kill bacteria. These include: cloves, lemongrass, bay leaf, chili or cayenne pepper, rosemary, marjoram, mustard, lime, garlic, onion, allspice, oregano, thyme, cinnamon, tarragon, cumin, caraway, mint, sage, fennel, coriander, dill, nutmeg, parsley, cardamom, pepper, ginger, anise, celery seed, and lemon. Vinegar can kill E. coli bacteria. Raw honey can kill some staph germs.

596

... But Don't Pay For It With Money!
A doctor at the University of California found that 10% of dollar bills tested contained bacteria capable of causing diseases, especially food poisoning via E. coli and Staphylococcus.

597

University of Iowa researchers believe that the helminth worm might be introduced into the human bowel to actually help improve the immune system by improving the amount of normal intestinal bacteria. It is possible this may help people with inflammatory bowel diseases and ulcerative colitis.

598

Doctors advise pet owners to deworm their pets regularly to reduce the chance of catching a bug from them. Wash your hands after handling pets. Keep pets out of the bedroom, and especially, off of the bed.

599

By September 2003, the West Nile Virus was off to a rip-roaring start as mosquito season got underway. Cases in humans had been reported in 17 states, versus four states for the previous year. Interesting changes included more infections in horses, a younger average age of infected humans (45 versus 55), and more people infected having encephalitis or meningitis.

600

Germs win, hands down! The average germ count on one square centimeter of the average male: thumbnail, 50-900 million; index nail, 800,000-1 million; other fingernails, 100-1.2 million; palm of hand, 100-5,000; back of hand, 400-1,000. For women: thumbnail, 350,000-650,000; index nail, 850,000-17 million!; other fingernails, 250,000-700,000; palm, 450-2 million; back of hand, 25-200. Hmm, next time someone says, "Let's shake on it" . . . maybe not.

601

University of Rochester researchers have proven that Streptococcus mutans (the bacterium which causes tooth decay) can travel from the mouth to the bloodstream and lodge in the heart, creating the danger of a life-threatening infection.

602

Scientists report that chiggers (a tiny Southeast Asian insect that loves to chow down on your legs and ankles) might provide the key to a low-cost treatment for AIDS.

603

Doggone!

Kennel cough is an infectious disease of the airways. Dogs and cats are most susceptible and the Bordetella bronchiseptica germs that cause the disease are transmitted by direct contact between animals, in a kennel setting for example. Kennel cough can spread to humans, though it is rare. Symptoms are similar to a cold or bronchitis. Whooping cough is caused by a related bacteria, but does not spread from pets to people.

Hi Yo!: Silver kills all known forms of bacteria, virus, and fungus.

604

One convoluted theory raises the premise that HIV-causing AIDS did not come from infected African chimps. Instead, some proponents say, HIV may have first come from a pig intentionally infected with African swine fever virus in Cuba in the 1970s when America attempted to cripple the country's pork industry. Supposedly, Haitian immigrants who had worked processing pigs in Cuba might have brought the disease to the U. S.

605

Germ Breath?

Millions of people suffer from chronic halitosis—bad breath. The source for this offending odor is germs. The tongue serves as a breeding ground for bacteria, which create sulfides which cause the stinky smell. Tooth-brushing reduces mouth odors by 25%. Tongue-scraping removes germs and food debris which feed the germs, reducing odors by 75%. Brushing and scraping reduce bad breath by 85%.

606

Cats with ringworm can be "healthy carriers," carrying and spreading disease while showing no sign of infection themselves.

607

A New York study showed that people, such as college students, who wash their clothes in washing machines used by others are more likely to get diarrhea or the common cold. To avoid this, it's recommended that you run the washer with 1/2 to 1 cup of bleach and hot water, then wash and rinse before you add your clothes.

608

German priest, Peter of Prague, broke bread for communion at the Church of Saint Christina in Solsena, Italy. While it must have seemed like a miracle when the broken bread had blood on it, scientists suggest that it was a red bacterium, Serratia marscesens, known to grow on bread stored in a damp place—and also a potential bioterror agent!

609

Shades of the Holocaust: In 1932, the U.S. Public Health Service began a study at the Tuskegee Institute in Alabama to determine the effects of untreated syphilis. Until 1972, almost 400 poor black men from the Macon County area were intentionally not treated for their latent syphilis. The uneducated men were not told that they had tested positive for syphilis, nor that they were part of a study. It's possible that as many as 100 of the men died from the effects of their untreated disease. Once the public learned of the experiment, it was discontinued. In 1974, the survivors were awarded a mere $37,500 each in an out-of-court settlement. A formal apology to the men was not given until 1997.

610

If you car smells funny right after you turn on your air conditioning, it could be caused by bacteria and fungi trapped in the system. You can have the system flushed with an antimicrobial, and let the engine run for a few minutes with only the ventilator on to dry up trapped moisture that provides a breeding ground for germs.

What you don't know about germs could hurt you!

611

The Ebola virus is a filovirus. Its natural host in the wild is unknown. It was discovered in 1976 in newly developed jungles when an outbreak in Zaire killed 90% of those infected. Ebola comes in 5 strains, each named for the place where it was first documented: Ebola Zaire; Ebola Sudan; Ebola Reston (Virginia); Ebola Tai; and Ebola Ivory Coast. Ebola is a hemorrhagic disease which eventually causes those infected to have their cells so destroyed that blood leaks through them causing conditions such as bright red whites of the eyes. As the diseases advances, it can cause bleeding from the nose, mouth, ears, rectum, and vagina. There is no cure.

612

By late summer 2003, only about 38,000 health care workers nationwide had been vaccinated against smallpox. Only nine states had more than 1,000 people immunized: Texas, with 4,241; Florida, 3,791; Tennessee, 2,429; Ohio, 1,772; California, 1,599; Minnesota, 1,476; North Carolina, 1,275; Missouri, 1,253; Louisiana, 1,107. States and major cities with less than 100 people vaccinated: Alaska, 96; Chicago, 72; Maine, 63; Arizona, 39; Rhode Island, 35; Nevada, 17.

Learn more about germs in
The Official Guide to Germs!

613

SARS Stuff: A U.S. strain of SARS was found to be milder that the Asian version of this disease. It was designated MARS. Advertisements to lure tourists to Hong Kong, saying "Hong Kong will take your breath away!" were quickly replaced following the outbreak of the SARS epidemic there. Epidemiologists fear SARS may never be eradicated among China's 1.3 billion citizens.

614

Mummy's Curse or Germ Curse?

In 1922, the Egyptian tomb of Tutankhamen, the famous boy king, was opened. An inscription on the tomb swore, "Death shall come on swift wings to him that toucheth the tomb of Pharaoh." Shortly thereafter, the excavation's benefactor, Lord Carnarvon, died from unusual complications following a mosquito bite. Within six years, at least 21 others who had some connection to the opening of the "cursed" tomb died of unnatural causes. Curse? In recent times, scientists have discovered that when such a tomb is first opened, mold spores—which can survive thousands of years—can burst out in an invisible cloud and enter the eyes, nose, or mouth of those nearby. Some molds are so deadly that even a small dose can kill you. Today, archaeologists wear protective masks and gloves when unwrapping mummies.

In the ever-, ever-, never-, never-land of germs, it just may prove to be that Mummy was always right. Keep your finger out of your nose, wash your hands, etc.!

A Flabbergasting Germs Glossary

Amerithrax: government term for the 2001 anthrax attacks; their investigation was named Major Case 184, with two investigative teams called Amerithrax 1 and Amerithrax 2

Ames strain: type of strain (collected from a dead cow in Texas in 1981) found in the anthrax attack letters of 2001; a natural form of anthrax

amplification: the replication or multiplication of a virus

"animalcules": what the first germs seen under a microscope were called

Animal Efficacy Rule: says that since it is illegal to infect a human with an infectious disease such as smallpox in the name of research, a drug or vaccine can be tested in two different animals with the disease, if the disease resembles the human form of the disease

antibiotics: strictly speaking, these are substances found only in nature

antimicrobials: includes natural and man-made antibiotics

"anxious face of smallpox": description used to indicate the common symptom of a worried look on the faces of those infected prior to the breaking out of pox pustules on the skin

archeobacteria: prehistoric species of bacteria

astrobiology: the study of germs on other planets

Bacchus: secret CIA program to build a small anthrax bioproduction plant to see if others could do the same with ordinary equipment available in the average hardware store; they could

Bacillus thuringensis (BT): anthrax that only affects insects; used by gardeners to kill grubs; the Iraqis once claimed that their Al Hakm anthrax facility was used strictly to create anthrax to kill grubs

bacteria: one-cell organisms; the smallest creatures on the planet

bacteriocidal: any drug that kills a bacteria

bacteriostatic: any drug that keeps bacteria from replicating

bifurcated: the twin-pronged needle used to give a vaccination against smallpox

"big pharma": term meaning large drug companies

Big Red Book: nickname for the book *Smallpox and Its Eradication*

biofilm: the invisible stuff (germs, fecal material, blood, etc.) that coats items such as IV and urinary catheters, prostheses, and endotracheal or gastric tubes

Biopreparat: the Russian laboratory that is home to half the world's supply of smallpox, as well as many other lethal germs as well as that country's bioweapons plant; also known as the System

Black 6: a purebred laboratory mouse naturally resistant to mousepox

black patents: nickname for classified patents given for the making of bioweapons in the U.S.

blebs: cold sores

"Blue Suit": Chemturion bright blue pressurized, heavy-duty biological space suit worn in Biosafety Level 4 areas to conduct work on deadly viruses

blue-suit work: working in a Level 4 lab, or working with Level 4 microbes

bricks: shape of smallpox under an electron microscope

broad-spectrum: drugs that are effective against a wide variety of diseases

BWC: Biological Weapons and Toxin Convention; banned the development, possession, or use of biological weapons; signed by more than 140 nations; some abide by this convention, others do not

carriers: people who carry a disease and can pass it on, but suffer no symptoms or ill effects themselves

centrifugal rash: the typical symptom of smallpox where the rash tends to cluster most thickly on the face and then outward to the extremities of the body

children's scissors: the only type (because they are not sharp) allowed in Level 4 laboratories

cold side: the area outside the dressing area of a Level 4 lab

commensals: bacteria that live on or in humans, usually keeping them healthy

community-acquired: disease-causing germs contracted outside of a hospital setting

compassionate use: permission given to use an experimental drug on a patient who has run out of all other options

Corpus 6: site of the official repository of one of only two places where live smallpox is held in freezers; this is at Vector in Siberia; the other is the CDC in the United States

"crash and bleed out": a term doctors use to mean that the patient's blood vessels have so completely eroded that the patient bleeds to death

"daughters": what scientists call the offspring of any bacteria

debridement: the surgical removal of dead or dying tissue

"dogbone of pox": term given for the dumbbell-shape core of a poxvirus

DOT (directly observed therapy): when patients must come into a clinic where health care workers can give out medication and see that it is actually taken

Dumbbell 7124: strain of smallpox; also called India, the strain which was the first to kill any animal

ectromelia: mousepox

emergence: the breakout of a disease from its source into the human population; sometimes the disease disappears so quickly that the source cannot be determined

endemic: a disease common in a particular geographic area

energetic: term to indicate that a bioweapon, such as anthrax, has been designed to bounce around so that it spreads more readily; a "professional" can create this; it is also called a "trick"

enteric: germs that live in the gut

Enterocytozoon bineusi: protozoa that causes chronic diarrhea in AIDS patients as well as in healthy travelers

EnviroChem: green liquid disinfectant used in chemical showers to kill viruses on medical workers' clothing

epidemiologist: a medical detective who tracks outbreaks to their source cause in hopes of stopping the epidemic and preventing future ones

The Eradication: common term for the efforts that put an end to smallpox in humans worldwide

exsanguination: a near-total loss of blood

"fecal carriage": medical term meaning the transfer of feces from person to person on the hands or other ways

flora: the germs, good and bad, that all people have on and in their bodies

foetor: old-fashioned term for an odor created when gas is given off from something like smallpox blisters; this stinky smell can help doctors diagnose the disease

Fort Detrick, Maryland: center of the U.S. Army's germ weapons research and development until 1969 when President Richard Nixon shut down all American bio-warfare programs

"gain of function": what geneticists call the amazing way genes can create new combinations between themselves

"germ theory": Louis Pasteur's concept that infections are caused by unseen microorganisms

GM: genetically modified

Gram negative: bacteria that turn a blue-purple color when treated with a stain (named for Danish doctor Christian Gram), then become transparent again when the dye is removed; these have two-layer cell membranes and include E. coli and other germs

Gram positive: bacteria that turn a blue-purple color when treated with a stain, and stay that color even if an attempt to remove the dye is made; these type of bacteria have single-layer cell walls, and include staph, strep, and enterococcus germs

gray area or gray zone: term for where the normal germ world and the world of hot zone infections meet; can be as large as a geographic area or as small as a room

gray room: staging area between the cold side and the hot side of a Level 4 lab; this is where you don your suits and gloves, or where you get disinfected upon leaving the hot side

"hamburger bug": nickname for E. coli because of its association with rare or undercooked ground beef

Harper: name (after the American soldier it came from) of a quite virulent strain of smallpox kept at the CDC

"hatbox": nickname for a cylindrical container made of waxed cardboard used to hold biohazardous material such as infectious viruses; also called an "ice cream container"

"heating up": the process of artificially speeding up the development of drug-resistant germs by those trying to create an especially lethal form of an infectious disease for the purpose of bioterrorism

hemorrhagic: diseases which cause such severe internal bleeding that they virtually dissolve organs

"herd immunity": when most of a population has immunity against a particular germ

hood: laminar-flow hood; laboratory cabinet like a kitchen stove exhaust which keeps samples from being contaminated and researchers from being infected

hot agents: highly-infectious germs which can cause diseases that may have no cure

hot side: term for the area of a Level 4 lab where you work with infectious agents

hot zone: term for a place (lab, room, building, village, etc.) contaminated with some type of biological "bug" that is highly contagious

HMRU: Hazardous Materials Response Unit, Federal Bureau of Investigation, FBI Academy, Quantico, Virginia

index case: the first person known to have contracted an infectious disease in an epidemic

in vitro: in the test tube or in the lab

in vivo: in humans or in other lifeforms

Lucy: nickname of 36-year-old overweight woman whose frozen corpse was dug up in Alaska to try to find live specimens of the 1918 Spanish Flu virus; these germs survived in Lucy because her extra fat had protected them from the ravages of time and frost

lyse: the term for the breaking down and shedding of bacterial cell walls

malaise: a general feeling of illness

MCL: Maximum Containment Laboratory; site in the CDC where one of the only two official stashes of smallpox is held in a freezer

mobile embalming pump: can be used to disinfect someone who died of an infectious disease

Monkey Cabinet: a large, portable aerosol chamber used to hold monkeys infected with infectious diseases

"mould juice": what Alexander Fleming called the first antibiotic, penicillin, which he discovered in 1928

MRSA: methicillin resistant S. aureus

"mulberry of pox": the term given for the bumpy, hand grenade-look (like the mulberry fruit) of a smallpox "pox"

narrow-spectrum: drugs that are effective only against a limited number of diseases

nosocomial: infections you get while in the hospital "getting well"

nuclear pox: a genetically engineered vaccine resistant smallpox virus; a.k.a. your worst nightmare!

"opportunistic infections": germs that otherwise might not bother a healthy person often wreak havoc in a person whose immune system is weakened because of other medical problems like HIV or treatments such as chemotherapy

"Orange Suit": bright orange, portable, pressurized space suit with battery-powered air supply; used in the field when working with airborne biohazards

panic button: red button on the wall where the CDC's smallpox freezers are kept; pushing it brings armed guards

panresistant: resistant to every drug there is

pathogens: germs that cause disease

Patient Zero: the first person to come down with a disease in an outbreak

"penicillin girls": once manually purified "mould juice" into penicillin

phages: viruses that kill bacteria

population analysis: when researchers consider the effects of antibiotics on a large sampling of bacteria

prophylaxis: something used as a prevention

pus: yellowish gunk found in infected tissue; consists of white blood cells, cell debris, and necrotic (dead) tissue

pustulate: the creation of pus-filled boils or blisters on the skin

the Rahima: name (after the girl they came from) for the smallpox scabs saved near the end of eradication; they still exist

recombinant: viruses that have been engineered or "constructed" in a laboratory; also called a designer virus because it is created to achieve a particular purpose

red man's syndrome: a hypersensitive reaction to a drug where the skin literally turns red

reverse isolation room: a contained area where fresh air is circulated into, then out, through a high-efficiency filter than can remove particles as small as bacteria

ring containment: the emergency procedure where people around the original index case of an epidemic are immunized in hopes of containing an epidemic to a limited area; a failed ring vaccination is called a containment failure

R-zero: term used to indicate the average number of new people who catch an infectious disease from each infectious person; also known as the multiplier of the disease, it is an indication of how fast a disease will spread

Salmonella typhimurium DT 104: causes food poisoning; highly resistant to antibiotics; spreads from cattle to people

SIDA: the French term for AIDS

signature: things scientists look for in bioweapons to determine which person, country, or lab might have created them

silica nanopowder: superfine powdered glass; used primarily for industrial purposes; such glass was found in the 2001 anthrax attacks spores

skull anthrax: term for a weapons-grade form of anthrax which has a shape like a skull

Smallpox Key: the "key" that disarms and unlocks the smallpox freezers at the CDC

smallpox seeds: eraser-size lumps of frozen amplified smallpox goop

spores: the hard, little capsules that bacteria form when they are threatened, where they stay safely encapsulated until they have an opportunity to strike again

STLV: a bioreactor that you can grow human tissues in, where the architecture of the tissue is preserved

Streptococcus iniae: found in Toronto, Canada; infection from tilapia or St. Peter's fish; causes skin infections and fever

surrogate: a fake bioweapon used for the testing and development of a real bioweapon

"terminal cleaning": the extreme disinfecting of a highly-infectious area such as an isolation unit

transfection: the introduction of foreign DNA into living cells

transposons: tiny bits of DNA that can bring antibiotic resistant genes into a cell

USAMRIID: (pronounced you-sam-rid) United States Army Medical Research Institute of Infectious Diseases; also called Rid or the Institute; main biodefense lab in the U.S.

vaccine: comes from the Latin word for cow

vaccinia: the live poxvirus used in a smallpox vaccine

variola: medical term for smallpox; from the Latin for "blotchy pimples"; variola minor is a lesser form of the disease, also known as alastrim; variola major is the more severe and deadly form

variolation: an older term for vaccination

vector: the origin or source of an infectious disease

VECTOR: the Russian State Research Center of Virology and Biotechnology, in Siberia

virulence: how strong, powerful, or "hot" a germ is

virus: Latin for "poison"; a microscopic piece of protein-covered DNA or RNA so small it is not even a living, cellular organism

virus melt: term for the soup of germy goop smallpox eventually leaves behind

VISA: vancomycin intermediate-resistant S. aureus

"The Visitor:" nickname of a man who peeked through a crack in the door of the hospital room of a woman who had smallpox; he was told to leave and did so, but came down with smallpox in spite of a prior immunization, which had obviously worn off

WMD: weapons of mass destruction; the FBI has a Weapons of Mass Destruction Operations Unit

Flabbergasting Further Resources on Germs

AMA Family Medicine Guide,
AMA Guide to Family Symptoms,
AMA Book of First Aid and Emergency Care

Arrowsmith
by Sinclair Lewis.
The story of a doctor who treats his
patients with bacteriophages.

Awakenings, a film that depicts a worldwide
epidemic of encephalitis in the 1920s.

The Bad Bug Book
US Food and Drug Administration website
http://vm.cfsan.fda/gov

The Centers for Disease Control and Prevention
www.cdc.gov

*The Coming Plague: Newly Emerging Diseases in a
World Out of Balance*
by Laurie Garrett

*The Complete Idiot's Guide to Dangerous
Diseases and Epidemics*
by David Perlin, Ph.D, and Ann Cohen
The Demon in the Freezer
by Richard Preston

DISEASES, 8-volume series,
edited by Bryan Bunch, Grolier Educational,
Danbury, Connecticut

Epidemic
by Brian Ward

Epidemic! The World of Infectious Disease
edited by Rob DeSalle

Federation of American Scientists'
Biological Arms Control Program

*Flu: The Story of the Great Influenza Pandemic of
1918 and the Search for the Virus That Caused It*
by Gina Bari Kolata

Germs: Biological Weapons and America's Secret War

Hazardous Materials Response Unit,
FBI, Quantico, Virginia

The Hot Zone
by Richard Preston

in.fullcoverage.yahoo.com/fc/India/Antibiotics_and
_Microbiology/

Invisible Enemies: Stories of Infectious Diseases
by Jeanette Farrell

*Johns Hopkins Symptoms and Remedies:
The Complete Home Reference*

*Johns Hopkins University Center
for Civilian Biodefense Strategies*

*Killer Germs: Rogue Diseases
of the 21st Century*
by Pete Moore BSc, Ph.D.

*The Killers Within:
The Deadly Rise of Drug-Resistant Bacteria*
by Michael Shnayerson and Mark J. Plotkin

Level 4: Virus Hunters of the CDC
by Joseph B. McCormick, M.D. (former head of
the CDC's legendary "hot zone") and Susan
Fisher-Hoch, M.D.

Love in the Time of Cholera
by Gabriel Garcia Marquez, Nobel prize winner in literature. An eerie, yet fascinating look at life during a cholera epidemic when vacationers sailed down rivers filled with dead bodies.

Mayo Clinic Family Health Book
Medicine Man, 1992 movie starring Sean Connery

Modern Meat
by Orville Schell

National Institutes of Health, Bethesda, Maryland

The New Killer Diseases: How the Alarming Evolution of Mutant Germs Threaten Us All
by Elinor Levy and Mark Fischetti

Office of Public Health Emergency Preparedness & Response, Washington, DC

Outbreak, fictional story/movie based on an Ebola outbreak in California, starring Dustin Hoffman

"The Pied Piper of Hamelin." This legend of a German piper luring rats to their death dates from the time of the Black Death.

USAMRIID: U.S. Army Medical Research Institute
of Infectious Diseases,
Fort Detrick, Frederick, Maryland

*Plague: A Story of Science, Rivalry and the Scourge
That Won't Go Away*
by Edward Marriott
Secret Agents, The Menace of Emerging Infections
by Madeline Drexler

Vector—State Research Center of Virology and
Biotechnology, Novosibirsk, Siberia

*WHEN EVERY MOMENT COUNTS:
What You Need to Know About Bioterrorism* From
the Senate's Only Doctor
by Senator Bill Frist, M.D.

When Plague Strikes
by James Cross Giblin

World Health Organization, Geneva, Switzerland

www.endofpolio.org:
website of polio-related photographs

www.germwebsites.com

www.keepantibioticsworking.com

Year of Wonders: A Novel of the Plague
by Geraldine Brooks.
An English village chooses to isolate
itself for a year to protect neighboring
villages from the plague.

*An American Plague: The True and Terrifying Story of
the Yellow Feaver Epidemic of 1793*
by Jim Murphy

After consuming all of these flabbergasting facts, you've probably thought of your own interesting and unique experiences with germs. We all have a million of them! Now I'm not asking for gross germy photos, but please send me your best stories about germs… the funny stories, sad stories, rescue stories, "on the job" stories, and triumphant stories. The Gallopade Gang will enjoy reading all of your germy anecdotes!

Gallopade International
Attn: Germy Stories
665 Hwy. 74 South
Suite 600
Peachtree City, Georgia 30269

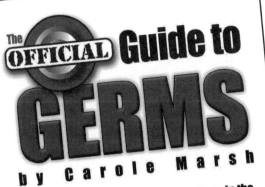